I

PALLADIAN STUDIES IN AMERICA

REGINA VIRTVS

BUILDING
BY THE BOOK
Second Annual
CONFERENCE AND TOUR
CENTER FOR PALLADIAN STUDIES
IN AMERICA
P. O. Box 5643, Charlottesville, VA 22905

IN RICHMOND
Virginia
28 - 30 May
1982

PALLADIAN STUDIES
IN AMERICA   I

BUILDING BY
THE BOOK
1

Edited by
Mario di Valmarana

PUBLISHED FOR THE CENTER FOR
PALLADIAN STUDIES IN AMERICA
BY THE UNIVERSITY PRESS
OF VIRGINIA, CHARLOTTESVILLE

THE UNIVERSITY PRESS OF VIRGINIA
Copyright © 1984 by the Rector and Visitors
of the University of Virginia
First published 1984
Library of Congress Cataloging in Publication Data
Main entry under title:

Building by the book.

(Palladian studies in America, ISSN 0741–7543 ; 1)
Papers delivered at the annual conference of the
Center for Palladian Studies, held in Richmond, Va., in
1982.
1. Neoclassicism (Architecture)—Virginia—Congresses.
2. Architecture, Modern—17th–18th centuries—Virginia—
Congresses. 3. Neoclassicism (Architecture)—United
States—Congresses. 4. Architecture, Modern—17th–18th
centuries—United States—Congresses. 5. Palladio,
Andrea, 1508–1580—Influence—Congresses. I. Di
Valmarana, Mario. II. Center for Palladian Studies in
American (Charlottesville, Va.) III. Series.
NA730.V8B84 1984   720'.975   84-2249
ISBN: 0-8139-1022-6 (v. 1)

To Frederick D. Nichols

FOREWORD

The Center for Palladian Studies in America in this volume presents to its readers the papers delivered at the annual Conference of the Center held in Richmond, Virginia, in 1982.

In order to promote the understanding of Palladianism, as in this issue, the publications that follow will contain material from future conferences planned in this country and Europe as well.

As president of the Center for Palladian Studies, I wish to express our gratitude to those who substantially aided the Center in carrying out this publication, especially Mario di Valmarana, editor of the series and vice president of the Center, and the anonymous Foundation whose generosity has made possible the printing of this volume.

Stanley Woodward
President

PREFACE

The subject on Palladian influence in eighteenth-century American architecture has been, for a long time, of great speculation among international scholars and critics alike. The exhibit of models of Palladio's buildings in Italy that toured this country from 1976 through 1978 stimulated a fresh rediscovery of Palladianism which later affected new directions in architectural thinking.

In 1979 I proposed to Frederick D. Nichols, then Cary D. Langhorne Professor of Architecture at the University of Virginia, and an authority on Jeffersonian architecture, to establish a center for the purpose of studying Palladianism. His enthusiastic support was indeed a determinant to organize a series of conferences, the second of which was held in Richmond, and from which this book comes. There, Freddy led a tour through the Virginia State Capitol, built by Jefferson, recounting the building's history and its critical implications for the symbolic architecture that ensued. The intimate, intense rapport that Freddy had with the structure, and the manner by which that rapport was conveyed to us, stimulated among us a great cultural moment and leads us to dedicate this volume to him.

I tender, for this, my deepest thanks to Freddy Nichols; and I am indebted to him for all his generous help he gave to me in carrying out this program.

I owe a great debt of gratitude, too, to the many kind people who have given so freely of their time toward the compilation of this book; in particular, I wish to thank Marlene Elizabeth Heck, whose knowledge in the field of eighteenth-century American architecture has been invaluable. To Joan Baxter, who has served as both typist and proofreader, I owe much for her skilled and willing assistance.

CONTENTS

PALLADIAN STUDIES IN AMERICA

I

BUILDING BY THE BOOK

1

WILLIAM HOWARD ADAMS

INTRODUCTION

Given the rapid spread of the British Empire throughout the world during the seventeenth and eighteenth centuries, it is not surprising that the printed book was the most efficient vehicle to carry its cultural baggage to the remote corners of the globe. The book was the backbone of education and cultural diffusion in the American colonies on every subject and in every field that it could legitimately translate.

Teachers trained in British schools, including the clergy, were important, of course, in bringing the classics, mathematics, law, philosophy, and theology to new generations of Americans scattered along the eastern seaboard on the western edge of European civilization. Young Thomas Jefferson's first encounter with men of learning, many of whom had studied in the cultural centers of London, Oxford, and Edinburgh, occurred in Williamsburg. As a student at William and Mary he had been taken in by a small circle of intellectuals led by Governor Francis Fauquier, William Small, and George Wythe.

Yet for all of their learning, the professional study of architecture was not a part of the background of the colonial intelligentsia. Though it was to become Jefferson's passionate obsession throughout his life, architecture was not accessible in the William and Mary classrooms or in the enlightened dinner table conversations at the Governor's Palace in the same way that the latest cases in Chancery could be analyzed or the advances in science could be critically discussed and debated. There is speculation that George Wythe's father-in-law Richard Taliaferro, who lived in Williamsburg, had studied architecture in England, and it is assumed that he may have discussed the subject with the enthusiastic squire from Albemarle County, but Jefferson never mentions it.

There is evidence that at least in the better houses of Williamsburg and along the Virginia rivers where the important building was taking place, the latest architectural fashions did receive attention. Thomas Tileston Waterman in *The Mansions of Virginia, 1706–1776*, provides numerous examples of the importation and adaptation of plans as well as design details taken from recently published English source books.

During the greatest period of plantation house building in Virginia, between 1720 and the Revolution, a number of the gentlemen builders had actually gone to school in England. Though they mainly pursued a classical or legal education, they were touched by the style of English country life that was sympathetic to the Virginia setting. If they were ambitious like Mann Page of Rosewell or Robert Beverley, the builder of Blandfield, the latest Anglo-Palladian architectural style impressed them as a model for their estates.

Without trained architects, an established tradition of architecture beyond a colonial vernacular interpretation of English building types, or an academic program of any kind, a Virginian with architectural pretensions and a well-chosen building site had only one place to turn for serious inspiration and guidance. The architectural treatise and pattern book would have had to be invented if only for the American colonies had it not already appeared some fifty years before the first British settlements were established in North America.

The first book on English architectural theory following Italian models was John Shute's *First and Chief Groundes of Architecture* published in 1563. Translations of Serlio, Vignola, and Palladio into English editions were to follow in the seventeenth century, but the English architectural book boom did not really get under way until the early eighteenth century.

The great Georgian building splurge in England, and echoed at least in the Virginia colony in an impressive way, was also well along by the 1720s. As Rudolf Wittkower has pointed out, the new books—from folios to journeymen's manuals—both appeared in response to the new building energy and at the same time was stimulated by it. A number of these books reached Virginia shortly after they appeared in London. There is evidence in several instances where published models were followed in the colonies within the year of first publication. When the library of William Byrd II, the builder of Westover, was catalogued in 1777 it could boast of twenty-seven architectural titles of both theory and practice. There were also a number of works on Roman antiquities. It is at this point that the participants in the second annual conference of the Center for Palladian Studies in America take up the story. The title, *Building by the Book*, aptly describes the fascinating interplay between the printed page and architectural development in eighteenth-century Virginia.

When Jefferson first began his self-taught studies of architecture in the 1760s, he also began to build the first of his three celebrated libraries. His passion for books prompted him to assemble one of the great private book collections in America. Over the years forty-nine books on architecture would be collected, in response to his unabated passion for building. But Jefferson was by no means unique in his recognition of the importance of the book and the printed image in the advancement of architectural skills and taste. He had, after all, proposed to George Washington to distribute Piranesi prints of classical buildings in Georgetown in order to lift the level of architectural expectation in the new capital. The chapters that follow present new evidence concerning our colonial ancestors' search for harmony and beauty following the rules and traditions of the art of building, a pursuit by no means yet concluded more than two hundred years later, even if the importance of the book in that search has been diminished, though not replaced. The hundreds of monographs on the work of individual architects in Europe, America, and Japan that appear each year bear witness to a well-established tradition.

LIONELLO PUPPI

PALLADIO, PALLADIANISM AND PALLADIANISTS 1570–1730

Translated by Joyce Vassallo Storey

In 1570 Palladio edited his *Four Books of Architecture*, prefaced by short and pertinent considerations on building techniques (figs. 1, 2, and 3). The treatise, the result of a long and complex preparation and conceived, in the words of the master, as the first part of a longer work, dealt with the five standard divisions: private houses, roads, bridges, squares, and temples. It was preceded by the publication in the same year of quires of books I–II and III–IV, which, though distinct and differently articulated, substantiate the material in the treatise. One version of it, filled with "many and beautiful things, which it would be a shame not to publish," was already known to Doni, who mentions it in the *Seconda Libraria* of 1555, saying that "the book is untitled, but from what one can learn from it, could be called 'the norms of true architecture.'" A year later, Daniele Barbaro agreed with Doni and added that Palladio had included in the work drawings and comments on the constructions he had planned and built. This could well be a draft very similar to—or perhaps the same one as—the one found in the manuscript fragments in the Museo Correr in Venice and the RIBA in London. Although proved by Eric Forssman to be in Palladio's son Silla's hand, these fragments substantiate the most salient points of Palladio's original manuscript, which Vasari probably saw during his stay in Venice in 1566 and which he freely used in writing Palladio's profile in the *Vite*. Palladio himself might have shown it to Vasari with the ulterior motive of strengthening an already favorable opinion in hopes of support for his subsequently successful petition for membership in the Florence Academy of Design. This was a very coveted honor to which Palladio truly aspired, while on the other hand he felt no obligation to accept the influential invitations of Emanuele Filiberto and Maximilian II. This honor would sanction the free-minded condition of the intellectual: Palladio had written his treatise with the precise intent of qualifying himself on a literary level, even if he was an "outsider" in the field. In fact, that which fundamentally differentiates Palladio's *Four Books* from the contemporary architectural treatises—and what, as we shall see, places the *Four Books* in a special perspective as reference works destined to everlasting influence and to a wide, and at times disconcerting, diffusion—is the preeminence given in book II to his personal experience. In this, Palladio proposes his intention to erase the day-to-day realities of his youth, menially spent in the craftsman's workshop, and sheds light on his response to the new name of Palladio given to him, one unforgettable day, by Giangiorgio Trissino (wouldn't we like to recreate it in our minds?). It also underlines his awareness of the mission that later the humanist from Vicenza, in his *Italia liberata dai Goti*, would sing in unrestrained celebration.

It is impossible not to notice in some sudden, unexpected, and revealing passages of the *Four Books* Palladio's deep, persistent conviction of his natural tendency for architecture and his insistence on professing the ethical, historical duty given to him by fate to restore "to a new interpretation" "the true beauty and grace of the ancients" under the guidance of Vitruvius, the "only ancient writer of this art." As a "man who was born not only to serve himself, but to benefit others," Palladio gave of himself with tireless devotion. But he was convinced that this gift must transcend, through the abstract, graphic self-image, the petty events of his daily labor at the work sites or the drafting desk. Thus he based the validity of his discourse on the perfect correspondence between his own predestined inclination for architecture (whose sole foundation was the heritage of the "ancient man, filled with complex knowledge") and the demands of a society perceived through its highest rank, the noble-spirited aristocracy.

A volume entitled *L'Italia nella sua gloria originari, rovina e rinascita*, by Edmund Warcupp, published in London in 1660 (but probably written a few decades earlier), deals with the "Revival," the "Resurrezione," of the "Original glory" of Italy. In this work Palladio, referred to as "famous" for his role as "Renovator," is praised as the exemplary representative of the "Revival" and as a cultural European model. In other words, in Italy, the "famous country *par excellence*," Palladio, according to Warcupp, was exemplary because he embodied and represented the architectonic notion of a precise, clear ideal of *nobilitas*.

We have reason to believe that Warcupp's familiarity with Palladio was through the *Four Books*, which Jacques Androuet Du Cerceau in 1572 was already having copied in his scriptorium, and which, a few years later, Alessandro Farnese transcribed in a series of drawings. Let us also remember that when the Englishman Inigo Jones arrived in the Veneto in 1610, some thirty years after Palladio's hurried burial rites at the end of that torrid August of 1580, he took along the *Four Books* and, when confronted with the live works of Palladio, wrote copious notes expressing his disconcertion over the difference he found between them and the drawings in the treatise. He wrote, "All of Palladio's works are lighter than in the drawings." *Light (svelto* in Italian) is a term which well expresses the difference, the distance he had felt. Several years later, La Teulier, director of the French Academy in Rome, wrote to Villacerf: "Although Palladio's book is well printed, his works, when viewed in the original, give a different impression." It was a polite but firm objection to Villacerf's letter dated October 7, 1697: "I don't understand how the trip to Venice can be useful for architecture. It seems to me that nobody goes to that country for this purpose. As for Palladio . . . his book is so well printed that it's worth the works *in loco*."

Here we can find the basic statement which allows us to examine specifically and concretely the genesis of Palladianism. The *Four Books* are the result of the cultural and mental processes in which Palladio engaged in an effort to reconcile his actual practice of architecture with his theory of architecture and in which he formulated a posteriori a graphic, abstract synthesis of the lived experience through the acknowledgment of antiquity. Such a synthesis takes place in the pages of the treatise and finds its realization in the rigid accomplishment of a predestined and accepted task. Consequently, the *Four Books* represent a single though dialectic event: the completion of the Palladian reality, not as the beginning of a pertinent critical tradition, but as a formula, a tool, a method to be used, hypothetically, in the most varied contexts sharing a common nature of classicistic order, thanks to the flexible but efficient applicability of the method and to the classical evidence of the proposed models.

From all this is born Palladianism, a phenomenon both literary and graphic in its origins. The history of this phenomenon coincides with the spread and circulation of the *Four Books* and does not require a cultivated knowledge or serious understanding of Palladio's actual architecture, whose direct experience is subordinate to the authority of the plates in the treatise (Jefferson, for instance, never saw Palladio's structures). Thus the spread in the intensity of American and European interest in Palladio was based on his treatise as a reference, and on its interpretations, which are abstract and hence separate from the confrontation with the actual structures and, as a rule, from the original drawings, both projects and surveys. The importance and meaning of the English reception of Palladio is yet to be examined and reevaluated. The case of Inigo Jones is unique in its own right, and its insertion in the continuum of the Palladian tradition is the result of a later project, begun at the suggestion of Giacomo Leoni and related to Colin Campbell's *Vitruvius Britannicus*, by Lord Burlington and his circle (fig. 4).

Meanwhile, in agreement with Kubler, the Spanish prelude of Palladianism should not be overlooked but rather further developed. The Iberian penetration of the *Four Books* a short time after the *editio princeps* (De Franceschi, 1570) was not a casual happenstance. Even less casual was the treatise's initial translation of 1578 by Juan de Ribero Roda, which lies unpublished among the collections of the National Library of Madrid, and the edition of 1625 of the *First Book of Architecture* published in Valladolid by the architect Francesco De Proves, a member of a family of master craftsmen linked to Juan de Herrera. Nor can the interest of Domenico Greco be forgotten. This interest, born in a restricted but not completely solitary intellectual climate, was accurately glossed by Barbaro in his *Vitruvio*. Greco grew to know and admire Palladio in Venice around 1572 and painted his portrait, which now hangs in the Royal Museum in Copenhagen. This and the canvas attributed to Magagnò, in the Valmarana family, are the only testimonies of the true image of the architect.

Careful examination, however, reveals that the origins of the spread circulation of the *Four Books* was French. The confirmation of this is the request by Thou to Paolo Gualdo in 1617 for a biography of Palladio (as well as, one should note, one of Trissino). Additional reinforcement is found in 1645 in the work of Le Muet, who used the first book as the standard interpretation of the orders.

Totally different is the enterprise of Roland Fréart de Chambray. A friend of Poussin, he was a leader of the Society of the Intelligents which was organized around the superintendent of royal works, Sublet de Noyers, whose theoretical orientation led to the foundation of the Académie Royale d'Architecture in 1671. It is this interest in Palladio's treatise which explains the antimannerism directed against the "abus qui se sont introduit dans l'architecture" and which can be linked to the resumption of work on the Louvre, the monumental center of the classicistic cultural tendencies of Sublet de Noyers. Hence we find the utilization of Palladio in an antimanneristic and antibaroque polemic, by means of, once again, the reduced and purified abstraction in the *Treatise*.

The research by Fréart in Venice, however interesting it may be with regard to the transmission of the graphic history of the drawings (a complex subject still to be explored), should not mislead us. This research focused exclusively on plates prepared for books on architecture, which were to have been printed but never were. For his edition, in fact, the Frenchman came in possession of four drawings, which he published—rather than use the original available plates of 1570, except for the frontispiece—so that his efforts did not involve the identity of the documents related to Palladio's study of ancient buildings.

Coming back to our main point, the French initiative was joined by Georg Andreas Böckler's German version of the first two books of Palladio's treatise (planned during the second half of the sixteenth century), published by Johan Andreas Endeter, Nuremberg, 1698, with a rather impressive title *Die Baumasterin Pallas oder in Teutschland erstandene Palladius*. While the illustrations convey the style of the original engravings taking into account the manneristic Rudolphian influence, the frontispiece plate illustrates a classicistic vision of the Rotonda as the privileged architecture of the villa and was engraved by Glotsch from a drawing by Johann Jacob von Sandvart, closely connected to the program of the Teutsche Akademie. Then, in 1694, when the Akademie printed the second part of the *Palazzi di Roma*, it attempted to integrate that work with a sequence of illustrations (not well known to scholars) of "Praedia aedesque hortenses immortaliis gloriae architecti Andreae Palladii: iussu ipsius hinc hinde in Statu Veneto a fundamentis erectae" based on the plates of the *Four Books*. It was upon this foundation that the Venetian Giacomo

Leoni appeared, on the invitation of the Palatine elector (who was of classicistic and antirococo inclinations), and prepared in 1708, from Palladio's treatise, his interpretation of the *cinque ordini* that would later be developed in England, when he moved to London shortly thereafter.

As Rudolf Wittkower noted, just as Leoni moved to London Pellegrini moved from London to Düsseldorf. This spectacular exchange of domiciles is not for us to speculate on, nor is it easy to explain. It is certain that Leoni was in England in the personal service of Henry, duke of Kent. And we now have a corroboration of Palladian influence on the basis of drawings of 1715 found by Timothy Hudson in the Bedfordshire Record Office relating to the building version of a house for a nobleman and from a manuscript entitled *Compendious Directions for Building*. Both, first of all, acknowledge Kent, their patron and Maecenas, and his ability as "judge of the noble art of architecture," and both suggest a well-established manner of building domestic architecture.

Yet we should ask if the sudden move of Leoni, well ensconced until then at the Palatine court, had reasons other than the simple assumption of services to the duke of Kent. In other words, can a connection be made between Britain's rapidly changing political and cultural situation and a systematic search for new personnel, artists and workmen, who could produce new monuments quickly and systematically? This was the time in England when the bankruptcy of Tory politics was confirmed by the transition from the Tories and the Stuarts to the Whigs and the Hanoverians in 1713-15, which, consciously or not, led to the elimination of the artistic baggage of the preceding dynasty. It was the period of the governing oligarchy surrounding the German-speaking and thoroughly un-English George I: the Whigs, the liberal democrats, and the moderates. These men became the partisans of a new and austere aesthetic, both antibaroque and classicistic, which vigorously emerged from the last and travailed years of Anne's reign. It was not by coincidence that in 1715 the first volume of the monumental *Vitruvius Britannicus* was published. The dedication is to Colin Campbell, "Doctor of Laws," but the reference to "Antique Simplicity" is connected with the "renowned Palladio" and the illustrations were clearly inspired by the Italian architect's treatise, derived, it has been hypothesized, from studies of Italian buildings *in loco* by James Smith, Campbell's principal graphic artist.

In fact, I am convinced that the success of the *Four Books* in England had been modest up to that time, and that the influence of Palladio was only associated with sojourns into Italy on "Grand Tours," where his treatise served as a memorandum and guide. This is not to deny, however, as noted by Sir John Summerson, a few examples of Palladio's influence on Elizabethan architecture and, of course, the works of Inigo Jones. It was because of Godfrey Richard's 1663 publication, in English and with various heterogeneous additions, of the *First Book* based on the 1645 version by Le Muet that there were timid hints of classicism in the empirical and scholastic tones of the work, rendered all the more visible in 1664 in John Evelyn's translation of the *Parallele* by Fréart de Chambray. And it is no coincidence that the two works were reprinted respectively in 1708 and 1707, the years in which the Englishman Thomas Twesden, in Vicenza, was trying to attract Francesco Muttoni to England.

If my arguments are correct, Leoni's sudden move to England was the consequence of an invitation deliberately extended to the major Palladian expert in order to provide British architectural culture with the canonical text of a true classical order, on the example of Fréart de Chambray (whose works had served as a source for Leoni). More particularly, he would impose a solemn and perfect monumental order that would appear to be connected with and assimilated to a national historical process: from the very beginning Leoni promised to publish Inigo Jones's Palladian glosses, which he was only able to give to the public in 1742.

Let us consider a few seemingly unconnected but nonetheless significant facts. The first volume of Leoni's edition of Palladio's work, bearing the date 1715, appeared in 1716 (in the light of Wittkower's accurate analysis) (fig. 5). This first volume must have required a certain amount of preparation, whether (1) for the redrawing of the plates or (2) for the work's organization by the publishers, the translators, and the financial backers of an oeuvre which, by all rights, the author prefatorily boasted, was "*more beautifully* printed [than] in Europe." There is no doubt that upon his arrival in London, Leoni availed himself of graphic materials already well organized and prepared, and in his preface he alludes to his own five years of continuous work (which Nicholas Dubois, the translator into French, more cautiously considered to have been more than a "few years"). As for the rest, the story is a different one. We know that Sebastiano Ricci, then available in London, was commissioned to prepare the allegorical frontispiece and Palladio's portrait and the overseeing of the English engravers was delegated to a person of international reputation: the Dutchman Picart. All this meant time and a lot of money, the latter of which was supplied by a group of underwriters whose names appear in lists attached to the first, third, and fourth volumes. If a sociopolitical and cultural examination of these names were undertaken (and it has yet to be done), it is very unlikely that a foreigner, "generous" and courageous but nonetheless an anonymous newcomer to England, would be able to put together such a group. The likelihood that Leoni arrived by invitation seems reinforced. The close collaboration from the very beginning between the Venetian and Dubois, a confidant of the duke of Marlborough, is to be considered. The close relations maintained by George I with the German world from which he came (the Palatine dynasty was consanguine with that of Hanover) are also not without their significance.

The work, in its totality, is dedicated to the English sovereign. "When and where" exclaims Leoni, "could Andrea Palladio's perfectly written and delineated lessons in this Art so magnificently again see their true light—if not under the munificent reign of His Honorable Majesty? And such a noble undertaking, encouraged by the intellectual generosity of the Splendid British Nation, to whom one must not pay tribute if not to Him who is the most worthy, that is to say, to their Sovereign?" Here can be discerned the systematized concept which embodied the decisive, historical change that was discussed above and that Ricci's frontispiece solemnly allegorizes. The second book is dedicated to the Palatine elector, under whose patronage Leoni states that he furthered his Palladian vocation. The third book is dedicated to Charles I, landgrave of Hesse and very close to the house of Hanover, whose love of Palladian architecture Leoni praises. Thus, the existence of a continuum is transparently suggested and traced to the very origins of its transmission. It is not a question of Leoni's personal and cultural history alone. One must consider this phenomenon also as a chapter which created the fertile era of Palladianism at a moment that proved to be favorable and opportune throughout Europe and that resulted in crystallizing forever its foundation in the abstract graphics of the *Four Books*.

Giacomo Leoni remains pivotal to the decisive change in the process of understanding Palladianism even before the spectacular editorial achievements of Francesco Muttoni in the Veneto (fig. 6) (as preannounced by the Englishman Twesden), of Bertotti Scamozzi (preceded by Consul Smith and preferred by a still more indigenous Englishman, Peter Edwards), and the decisive Palladian conversion of Lord Burlington. In fact, Burlington's "conversion" to Palladianism no doubt resulted from stimulation and enthusiasm inspired by the appearance of the first two or three books of Leoni's series. After Burlington's general and rather touristic Italian excursion in 1714–15, he made his great trip in 1719—which would result in the transfer of the priceless treasure of the almost complete collection of Palladio's signed manuscripts to England and the publication in 1730 of the drawings of the monument of the *Romans' Baths* (*Terme dei Romani*). Paradoxically, Burlington's conversion, which Leoni, editing Alberti's artistic corpus, praised as being "of exquisite genius," "magnificent," and in "good taste," after a decade of editorial successes was determinant to the decline of Leoni's fame, due to Burlington's precise philological requirements which led him to question the validity of the plates in Leoni's 1716–20 edition of Palladio. Burlington, using that same Campbell to whose name, according to Dubois, Leoni had wanted to link his own name, had already succeeded in putting together an English edition of Palladio's first book in 1728. Another new edition of the *Four Books*, this done by Isaac Ware and containing slighting remarks about Leoni's edition, had been out for four years when the Venetian, after a second edition of 1721, issued with modest success his 1742 reprint, enlarged with the transcription of Inigo Jones's notes and the translation of *L'antichità di Roma* (*Rome's Antiquity*).

American Palladianism derives directly from English Palladianism. Above all, it reflects, or rather, intensifies, English Palladianism's preference for the concept of the villa. In fact, texts and graphics from Leoni to Burlington had established in England a rich heritage of country house architecture.

The transfer of British Palladianism to American Palladianism remains, once again, an abstract and textual phenomenon. There are the great editorial monuments of the English Palladio, particularly those produced or published by Lord Burlington, which from the very beginning were circulated, accepted, and used in the British colonial world of the North American Atlantic seaboard. But out of this first phase of acclimation grew a conscious attempt to simplify and codify these reference tools in order to respond, in terms reflective of a simple but effective functionality, to the economic realities of the plantation system of Virginia, Maryland, South Carolina, and some other colonies, resulting in the motif of the villa as the governing nucleus of a farming concern and the comfortable residential space that fit the requirements of an emerging landed aristocracy. Out of this is born a fertile and extraordinary literature which included, from the very beginning, volumes in folio, but which was characterized in the end by the reduction of his material into manuals of an immediate and practical scope designed to seek out the widest approval and application on the part of the colonists. The diffusion of the graphic representation of Palladio's architectonic typology, which had become the preferred testimony to Palladio's art and which was trimmed down to the minimal skeleton of its functional structures and its representation of a significant, classicized image, i.e., the villa, provided the basis for an extraordinary process of architectonic development. This development, of course, did not exclude the recovery of other motifs provided in Palladio's repertoire. We have only to consider the rediscovery of the acropolitan model, attested to in Richmond by Thomas Jefferson (fig. 7). The energy of this rediscovery, nonetheless, appears somewhat directed toward the finding in Palladio (and actually only in Palladio the treatise writer) of a methodology rather than a typology in order to introduce a lesson into a widely varied and fertile intellectual environment, an environment that was, however, shaped by the desire to create a harmonious balance between natural shape and man-made space. This desire is essentially Palladian.

Hence the American emergence and utilization of the Palladian graphic motif of the villa led to the construction, ordered adaptation, and development of an original and incomparable architectonic adventure. It was a manner of building capable of attaining both the constancy of clear and homogeneous themes and the emergence of stylistic variants—the former and the latter of a widely felt nature—in the setting of a historical dimension which was establishing, because of its detachment from the mother country, the parameters of its own civil and cultural autonomy. Both this constancy and this emergence of specific stylistic moments were the formal confirmation which panoramically marks and qualifies the immense expanses of the American South and which would result in, it may not seem excessive or out-of-place to affirm, Henry Hobson Richardson and Frank Lloyd Wright.

1. *Title page, Andrea Palladio,* The Four Books of Architecture *(Venice, 1570), bk. 3.*

DE I DISEGNI DELLE CASE DI VILLA DI ALCVNI
nobili Venetiani.   Cap. XIIII.

L A FABRICA, che fegue è in Bagnolo luogo due miglia lontano da Lonigo Ca
ftello del Vicentino, & è de' Magnifici Signori Conti Vittore, Marco, e Daniele fra
telli de' Pifani. Dall'vna, e l'altra parte del cortile ui fono le ftalle, le cantine, i gra
nari, e fimili altri luoghi per l'ufo della Villa. Le colonne de i portici fono di ordi
ne Dorico. La parte di mezo di quefta fabrica è per l'habitatione del Padrone: il
pauimento delle prime ftanze è alto da terra fette piedi: fotto ui fono le cucine, &
altri fimili luoghi per la famiglia. La Sala è in uolto alta quanto larga, e la metà più: à quefta altezza
giugne ancho il uolto delle loggie: Le ftanze fono in folaro alte quanto larghe: le maggiori fono lun
ghe un quadro e due terzi: le altre un quadro e mezo. Et è da auertirfi che non fi ha hauuto molta
confideratione nel metter le fcale minori in luogo, che habbiano lume viuo (come habbiamo ricor
dato nel primo libro) perche non hauendo effe à feruire, fe non à i luoghi di fotto, & à quelli di fopra,
i quali feruono per granari ouer mezati; fi ha hauuto rifguardo principalmente ad accommodar be
ne l'ordine di mezo: il quale è per l'habitatione del Padrone, e de' foreftieri: e le Scale, che a queft'or
dine portano; fono pofte in luogo attifsimo, come fi uede ne i difegni. E ciò farà detto ancho per
auertenza del prudente lettore per tutte le altre fabriche feguenti di un'ordine folo: percioche in
quelle, che ne hanno due belli, & ornati; ho curato che le Scale fiano lucide, e pofte in luoghi commo
di: e dico due; perche quello, che uà fotto terra per le cantine, e fimili ufi, e quello che uà nella parte
di fopra, e ferue per granari, e mezati non chiamo ordine principale, per non darfi all'habitatione de'
Gentil'huomini.

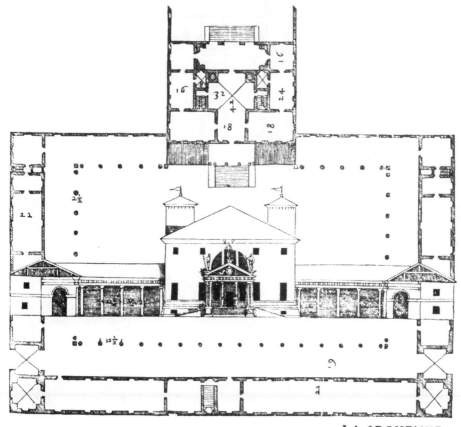

LA SEGVENTE

LA SEGVENTE fabrica è ſtata cominciata dal Conte Franceſco, e Conte Lodouico fratel-
li de' Triſsini a Meledo Villa del Vicentino .　Il ſito è belliſsimo : percioche è ſopra un colle, il quale
è bagnato da vn piaceuole fiumicello , & è nel mezo di vna molto ſpacioſa pianura , & à canto ha vna
aſſai frequenteſtrada.　Nella ſommità del colle ha da eſſerui la Sala ritonda , circondata dalle ſtanze,
e però tanto alta che pigli il lume ſopra di quelle . Sono nella Sala alcune meze colonne , che tolgo-
no ſuſo un poggiuolo, nel quale ſi entra per le ſtanze di ſopra ; le quali perche ſono alte ſolo ſette piedi , ſeruono per mezati . Sotto il piano delle prime ſtanze ui ſono le cucine , i tinelli , & altri luoghi.
E perche ciaſcuna faccia ha belliſsime uiſte ; ui uanno quattro loggie di ordine Corinthio : ſopra i
frontespicij delle quali ſorge la cupola della Sala . Le loggie, che tendono alla circonferenza fanno
vn gratiſsimo aſpetto : piu preſſo al piano ſono i fenili, le cantine, le ſtalle, i granari, i luoghi da Gaſtal-
do, & altre ſtanze per vſo di Villa : le colonne di queſti portici ſono di ordine Toſcano : ſopra il fiume
ne gli angoli del cortile ui ſono due colombare .

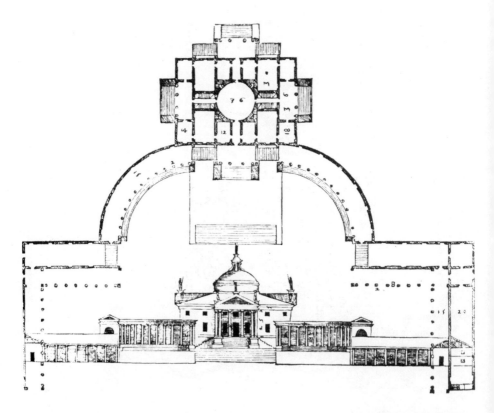

LA FABRICA

3.　*Villa Trissino, Andrea Palladio,* The Four Books of Architecture *(Venice, 1570),*
bk. 2, p. 60.

THE

DESIGNS

OF

INIGO JONES,

Confifting of

PLANS and ELEVATIONS

FOR

Publick and *Private* Buildings.

Publiſh'd by WILLIAM KENT,
With ſome Additional Deſigns.

The FIRST VOLUME.

M. DCC. XXVII.

4. *Title page,* The Designs of Inigo Jones, *ed. William Kent (London, 1727), vol. 1.*

5. *Frontispiece*, The Architecture of Andrea Palladio, *ed. Giacomo Leoni (London, 1716)*.

Æternitati.

TE MEMORANDE,
CANEMUS.
Vgr. Georg. 3.
M. D. CC. XL.

6. *Frontispiece,* The Architecture of Andrea Palladio, *ed. Francesco Muttoni (Venice, 1740), vol. 1.*

THOMAS JEFFERSON

ARCHITECT

ORIGINAL DESIGNS IN THE COLLECTION OF
THOMAS JEFFERSON COOLIDGE, JUNIOR
WITH AN ESSAY AND NOTES BY
FISKE KIMBALL

BOSTON

PRINTED FOR PRIVATE DISTRIBUTION

At The Riverside Press, Cambridge

M.DCCCC.XVI

7.   *Title page, Fiske Kimball,* Thomas Jefferson, Architect *(Cambridge, Mass., 1916).*

CALDER LOTH

PALLADIO IN SOUTHSIDE VIRGINIA: BRANDON AND BATTERSEA

Andrea Palladio hardly could have expected that his inge-
nious designs of villas and farmhouses published in his revolutionary *Four Books of Architecture* would be reinterpreted
two centuries later thousands of miles from Italy in a remote
corner of the earth, the English colony of Virginia. How and
why this New World wilderness should be peppered with
quite appealing versions of domestic concepts laid down by
a north Italian architect long before the Old Dominion was
ever dreamed of is the result of an interesting set of circumstances, particularly in the case of what are regarded as two
of the most refined and "Palladian" of Virginia's early houses,
Brandon and Battersea.

To see how these houses came about we should look at
what it is about Palladian designs that would have caught the
attention of Virginians in the first place. In his book of villa
and farmhouse designs Palladio clearly illustrated with plans,
elevations, and written description a revolutionary notion that
relatively modest country seats could be given architectural
style, could be made eminently commodious and practical,
and could relate directly and openly with the surrounding
landscape (fig. 1). No longer was a house to be a defensive,
inwardly turned castle; now it could be an open, outward-
looking dwelling in the modern sense, a building broken down

into pleasingly proportioned symmetrical and functional units, with a careful eye toward architectural effect.

Palladio's villa designs ranged from the excessively elaborate to the relatively plain, but they all possessed a straightforward formality and liveliness that was to have a particular appeal, surprisingly not to the fashionable set of continental Europe, but to the nobility and gentry of eighteenth-century England and its American colonies, especially Virginia. In the century following his death, Continental domestic design veered from the clarity of Palladio's classicism to the florid and convoluted baroque and rococo. It was the intellectual and artistic dilettanti of early eighteenth-century England, Lord Burlington and his circle, who recognized in Palladio an architectural attitude consistent with the tasteful rationalism of the Augustan Age. Thus adaptations, if not veritable copies, of Palladian villa designs became de rigueur for Britain's newest smart set. If Palladio had not the remotest connection with English history and traditions, no matter; his approach to building was certainly akin to the island kingdom's new intellectual spirit. Editions of Palladio's *Four Books* were mandatory for a gentleman's library, and English architects began churning out designs for villas and farmhouses that Palladio himself might have a difficult time recognizing as not his own. Indeed, the English countryside saw itself being dotted with quite handsome and knowingly designed Palladian-style houses. And the designs published by such architects as James Gibbs, Robert Morris, and Colin Campbell show a genuine inspiration from the master.

As expected, colonists looked to the mother country for virtually all matters of taste, and Virginia was no exception. Various gentlemen, architects, and builders in both Virginia and its sister colonies owned editions of Palladio's works as well as English pattern books illustrating Palladian-style designs. However, throughout the first half of the eighteenth century in Virginia any building that was not the most elementary vernacular shelter was a rarity. The few buildings that passed as architecture were provincial, conservative adaptations of the very plain Queen Anne and early Georgian idioms of the mother country. Williamsburg's very simple, albeit carefully proportioned, two-story, five-bay rectilinear Wythe house is the classic illustration. By the second half of the eighteenth century a few, and really a very few, homes of the most prosperous planters began to show an unmistakably Palladian influence: Mount Airy, Blandfield, and Mannsfield (destroyed), with their finely articulated center blocks, embracing hyphens, and connected dependencies are the notable examples. These large houses are, for the most part, massive and monumental and lack the more domestic villa aspect of the Palladian house type, which featured an emphasized two- or three-story center section with lower wings connected or not to hyphens and terminal dependencies. Mount Airy and Blandfield thus belong more to the Palladian interpretations of the English architect James Gibbs; it was the more down-to-earth, smaller-scale Palladian forms and the broken-up masses that appealed to other English architects such as Robert Morris, whose designs provided inspiration for scores of American dwellings.

Credit for popularizing the Morris versions of Palladian schemes in Virginia, and making them a design source for houses spread from tidewater to the mountains, and from the piedmont down into North Carolina and even Georgia, belongs, it would seem, largely to Thomas Jefferson. Like the English dilettanti, Jefferson was fascinated with Palladio's designs, as well as with his recordings of ancient buildings and his codification of classical architectural principles. From Palladio, Jefferson turned to his native Virginia and saw it as a land bereft of sophisticated specimens of architecture. He thus returned to Palladio as his "Bible" for a new order of buildings suitable for his individual-minded Virginians. Breaking the mass of buildings into aligned units, either of three, five, or even seven, was an aspect of Palladian farmhouse and villa designs that especially attracted Jefferson. For example, each of Palladio's schemes for the Villa Cornaro, the Villa Pojana, and the Villa Barbaro has a central three-bay, pedimented pavilion flanked by lower wings with hipped or gabled roofs, then with low hyphens and perhaps terminal wings beyond. The rectilinear center blocks of Mount Airy or Blandfield had too much of the massive Georgian formality that Jefferson wished to avoid.

The chief example of Jefferson's interpretation of this type of stretched-out house is the first version of Jefferson's own home, Monticello, designed in 1771 and built soon after (fig. 2). In this first form Jefferson designed Monticello as a two-story central pavilion fronted by a two-level portico of superimposed Doric and Ionic orders. The pavilion was flanked by a lower, hipped-roof wing of one story with an attic story above the main entablature. This main central section was in turn flanked by long, low esplanades under which the service areas were neatly concealed, all in the Palladian manner.

The first version of Monticello (before it was later remodeled and enlarged and given a more French aspect with its dome, as we know it today) is the earliest Palladian-type house with which Jefferson's association can be documented. For a variety of circumstances it would now seem that Jefferson was directly or indirectly responsible for a considerable number of Virginia's three-, five-, and seven-part Palladian-type houses in addition to Monticello. Definite attribution is complicated by the unfortunate fact that with the burning of Jefferson's childhood home, Shadwell, in 1770 virtually all his books, papers and architectural drawings up to that time were destroyed.[1] Thus it probably will be impossible to formally assign Jefferson's authorship of the Old Dominion's two most outstanding Palladian houses erected before that annoying fire, Brandon and Battersea, though with an examination of circumstances, Jefferson's involvement with these two houses seems a virtual certainty.

Before discussing Brandon and Battersea, note should be taken of one very interesting house that likely served as a prototype for the other two, a dwelling now known as Taze-well Hall that originally stood on the southern edge of Wil-liamsburg (fig. 3). Tazewell Hall, as shown in rough plans on both the "Frenchman's Map" of 1782 and the Mutual Assur-ance Society of Virginia policy of 1809, was originally built as a seven-part scheme paralleling plate 3 in Robet Morris's *Select Architecture* (fig. 4).[2] Only through these early docu-ments is the Palladian aspect of the house known, since the two-story terminal wings and the hyphens were removed, a two-level portico added, and the one-story wings raised to two stories around 1836 when the house was owned by the Tazewell family. Compounding its architectural complica-tions, the house was moved from its original location termi-nating South England Street to an adjacent side lot in 1908.[3] However, in its original form as a Palladian image interpreted by Robert Morris, it presented a strikingly different image from the resulting rather uninspired rectilinear house.

The precise construction date of Tazewell Hall is not known. However, the house was built for John Randolph, a promi-nent public figure who was a cousin to Thomas Jefferson through his mother. If the 1758–62 period usually attributed to Tazewell Hall is correct, Jefferson, who obviously was ac-quainted with his cousin through frequent visits to Williams-burg while attending the College of William and Mary, would surely have been familiar with the house. However, to credit Jefferson with the design, despite his known interest in Pal-ladian forms, might be stretching the case a bit. For although Jefferson was an early convert to Palladian architectural ideals, he was only in his teens when this singular dwelling is sup-posed to have been built. Indeed, it is possible that the young intellectual could have influenced the design of his prominent cousin's house, one of the colony's major dwellings of the period, but it seems more reasonable that a yet-unidentified architect-builder, also familiar with Robert Morris's then quite new book of designs, should receive the credit. If this is the case, the young and impressionable Jefferson would surely have been intrigued by such a singular dwelling, one that stood in marked contrast to the boxy buildings elsewhere in the colonial capital that were so unimpressive to him. Writing in 1785 in his *Notes on the State of Virginia*, Jefferson said of the college and hospital buildings in Williamsburg, the largest structures standing in the state at the time, that they were "rude, misshapen piles, which, but they have roofs, would be taken for brick-kilns." Only the Capitol, with its two-level portico of superimposed orders, albeit probably crudely proportioned, received any slight praise from Jefferson who said it "is tolerably just in its proportions and ornaments."[4]

Whoever was responsible for its design, one may now gain some idea of the original appearance of Tazewell Hall and its debt to Palladio through Robert Morris as a result of its removal from Williamsburg in recent years and its rebuilding in Newport News. Following its sale by the Colonial Williamsburg Foundation to Lewis A. McMurran, Jr., the house was rebuilt with the second stories of its wings removed and the hyphens and two-story terminal wings reconstructed. The careful reconstruction/restoration revealed a quite sophisticated Virginia version of a Palladian-style villa rendered in the indigenous materials of wood frame sheathed with beaded weatherboards and covered with a shake roof. Interestingly, architectural investigation revealed that the two-story center section of Tazewell Hall departs from the Morris plate prototype by not being divided into a stair hall and saloon, but instead is one grand two-story hall with an intermediate entablature supported on pilasters, thus obviating the need for stairs except in the terminal wings. The space was divided into two stories in the nineteenth century but was returned to its early appearance in the restoration.

Tazewell Hall, whether by Jefferson or not, is Virginia's first known full-blown instance of Morris-style Palladianism. While it was nearly contemporary with the more Gibbsian-style of Palladianism as revealed in Mount Airy and Blandfield, it heralded things to come. Mount Airy and Blandfield, impressive though they are with their solidly forthright center sections and embracing wings, are expressions of the mid-eighteenth century, signifying little future direction; the massive Georgian manor house, even softened by Palladian overtones, would not be the style of coming decades.

It is a pity that authorship of the designs for such intriguing, architecturally developed houses as Tazewell Hall, Brandon, Battersea, and other similar Palladian types of the period should remain undocumented, but in the absence or lack of discovery of records telling us precisely how the houses came about, educated speculation and circumstantial evidence remain the only realistic means by which we can figure why these houses look the way they do. As already noted, given his youth, it seems unlikely that Jefferson can be given full or even partial credit for the design of Tazewell Hall. Brandon and Battersea, on the other hand, make another case. The association of Jefferson with Brandon, the beautiful eighteenth-century seat of the Harrison family on the banks of the James River in Prince George County, rests on several factors (fig. 5). First, Brandon's seven-part elevation is, like that of Tazewell Hall, unquestionably patterned on Morris's plate 3 in *Select Architecture*,[5] a work familiar to Jefferson and in his possession by 1770.[6] Second, Jefferson was a good friend of Benjamin Harrison for whom his father, Nathaniel Harrison II, built the main part of the house, it is believed, on the occasion of his son's marriage in 1765 to Anne Randolph of Wilton, a cousin and acquaintance of Jefferson's. Indeed, an old Harrison family tradition holds that Jefferson served as a groomsman in Harrison's wedding and designed the house as a wedding present.[7] Third, as revealed in his own house, Monticello, it is obvious that Jefferson had a particular fascination with Morris's type of Palladianism.

It is interesting to note that the first stories of Brandon's terminal wings predate the rest of the house, as shown by the earlier style of brickwork used there—Flemish bond with glazed headers and rubbed brick at the corners and around the openings. The brickwork on the second stories, and indeed on the rest of the house, is very plain, merely Flemish bond with no glazed header pattern, no rubbed brick dressings, and no gauged-brick jack arches, all of which one would expect to find on an important plantation dwelling of the colonial period. One may thus speculate that the terminal wings were originally built as one-story dependencies by Nathaniel Harrison II sometime in the first half of the eighteenth century with the expectation that someday he would add a conventional detached rectilinear plantation house between them. Such a composition existed at Kingsmill in James City County, which had a squarish house positioned between two one-story dependencies set perpendicular to the residence. However, through the intervention of someone with a fascination for Palladianism, particularly Morris's plate 3, a conventional Virginia domestic complex never materialized, and a radically different scheme following the precedent of Tazewell Hall came into being. The terminal wings were pushed up to two stories, and the space between them was filled with a house consisting of a two-story center section flanked by one-story hipped-roof wings connected to the terminal wings by low one-story hyphens. The only real difference between the resulting Brandon and the Morris design is that the hyphens contain actual rooms while Morris showed them as walled areas, the left one being a stable yard and the adjacent terminal wing the stable. In Virginia the stable would never be attached to the residence but would be in a separate building. This points to the irony that a design in an English pattern book called for a house more modest than what it inspired to be built in the colonies. The reverse is usually the case.

Sophisticated though Brandon is, its red brick walls give it a decidedly colonial look, and the sophistication is diluted because the brickwork itself is so unrefined. The exterior would appear less provincial and closer to the Palladian ideal if the brickwork had been stuccoed. Morris did not specify surface materials, but the very lack of any treatment of the walls of his plate 3 design conveys the impression of smooth stone or stucco, which, of course, would be very Palladian. Indeed, plate 3 executed in red English-style brick laid up in Flemish bond would appear very alien to Palladio. That Brandon was intended to be stuccoed seems quite obvious when one considers that no attempt was made to match the original and later brickwork in the terminal wings and that the brickwork of the rest of the building is so plain, if not crude, and completely inconsistent with what would be the norm for a Virginia house of quality. In fact, it would be hard to find a great Virginia plantation house of the period with less attractive brickwork than Brandon's. A parallel case of intended stucco never being applied is Mount Airy, but there the walls are of unrefined stonework rather than brick.

The interior of Brandon shows no particularly Jeffersonian features except for a Chinese lattice stair railing and a heavy Doric entablature, both in the passage of the west wing. While neither feature is uniquely Jeffersonian, many of his known architectural works—Monticello, Barboursville, Farmington, and the University of Virginia among them—called for Chinese lattice railings juxtaposed with forthright classical ornament. (Palladio would, of course, have found Chinese flourishes on classical buildings beyond comprehension.) With Brandon, it is odd that its most interesting woodwork should be relegated to a wing; one can only assume that equal if not richer decoration of this type was in the center of the house before it was removed in an early nineteenth-century remodeling when the present Federal stair and its impressive colonnaded screen were installed. Moreover, this remodeling required the removal of a wall separating the stair hall from the saloon, an arrangement similar to that which was called for by Morris's plate 3, less two small rooms. Brandon's other principal rooms, the dining room and the parlor, retain their original woodwork, which is fairly standard but very handsome full paneling with pedimented chimneypieces. Except for the interior alterations noted, the substitution of the original sash with nineteenth-century sash, and the replacement of the original pedimented dwarf porticoes on the main entrances with the present flat-roof porticoes, Brandon has remained remarkably unaltered and stands today in a beautiful state of preservation, the focal point of an enchanting formal garden on the river side and a splendid romantic park on the land side. The setting varies from that specified by Morris for his plate 3, which called for one side of the house to face the sea and the other to be enclosed with a garden. Nevertheless, Brandon's close adherence to the English prototype testifies to the eminent suitability of Morris's ingeniously interpreted Palladian schemes for the life-style of colonial Virginians.

Southside Virginia's other most conspicuous example of colonial Palladianism is Battersea (fig. 6). Formerly in Dinwiddie County and now incorporated into the western edge of the city of Petersburg on the banks of the Appomattox River, this dwelling also makes a strong case for Jeffersonian authorship. In his meticulously researched *Mansions of Virginia* the celebrated Virginia architectural historian Thomas T. Waterman was the first to suggest that Jefferson be credited with the design of both Brandon and Battersea.[8] The attribution of Battersea to Jefferson follows very similar reasoning as that of Brandon. As in the case of Brandon, Jefferson was a good friend of the builder, John Banister, and was related to Banister's wife, Elizabeth Bland, through the Randolphs and maintained a close friendship with her as well. Jefferson also served in several public bodies with Banister. As is well known through his correspondence and many domestic designs, Jefferson was Virginia's preeminent gentleman architect and delighted in the opportunity to plan houses for his friends. Thus it would hardly be wild speculation to suggest that Jefferson provided a neat Palladian scheme for his friend John Banister to replace the family home that burned in the early 1760s while Banister was studying in England.[9] Indeed, no one else even remotely approaches the possibility of being the architect of such a scheme. The resulting edifice, even though it has undergone later alterations, retains the unpretentious but exquisitely refined formality characteristic of Palladio's own works.

Unlike Brandon, Battersea is not closely modeled after a specific design by either Palladio or Robert Morris. Rather it is an amalgam of several ideas of both architects, and in its scale, proportioning, and plan it comes across as being unmistakably American. Because of the later alterations, it is difficult to determine Battersea's original appearance precisely; however, it was built in a five-part Palladian scheme rather than the seven-part scheme employed at Brandon and Tazewell Hall. Morris illustrates nothing quite resembling the five-part format, and although the design for the Villa Barbaro provides Palladio's only close parallel for a five-part house, Battersea is obviously in the Palladian spirit. It consists of a two-story, pyramidal-roofed center section flanked by one-story gable-roofed wings that act as hyphens connecting the pedimented gabled-end terminal wings. As an accent for the center section there is a Roman pinecone finial at the apex of the roof rather than a pineapple as with Brandon. With this strictly formal sectional layout, each member carefully proportioned and balanced to complement its adjacent member, Battersea achieves the Palladian ideal of an architecturally sophisticated farmhouse adapted for the local life-style. The Palladian or Italian aspect of Battersea was noted by the marquis de Chastellux during his visit in 1781: "Mr. Banister's handsome country-house . . . is really worth seeing. It is decorated rather in the Italian than the English or American style, having three porticoes at the three principal entries."[10]

Battersea's Italian flavor is heightened by its stuccoed walls. Although it seems fairly certain that Brandon was intended to be stuccoed originally, this is not clear with Battersea. The present stucco is obviously very old, and the fact that Battersea struck Chastellux as looking Italian would suggest that the stuccoing existed at the time of his visit; the stuccoing would set the house apart from its more English-looking brick neighbors.

However, the stucco could just as easily date from the early nineteenth-century alterations and could have been required to cover the patches and scars in the brickwork caused by altering the openings. An examination of the interior paneled wainscoting shows what happened with these openings, and it is worth explaining these changes in order to show that Battersea's original bay arrangement was considerably more straightforward than at present. Most of the changes took place on the principal (south) facade. The south facade hyphens were originally two bays where there is now a single three-part window. On the facade of each terminal wing were also two bays, for which was substituted a single Palladian window. The arched top of the center section in each Palladian window is false, or not expressed on the interior, as it would interfere with the ceiling. The original bay arrangement is preserved on the north, or river, front, although most of the windows were enlarged in the remodeling. This change in window size required replacement of the surrounding trim, so that now most of the windows are framed with nineteenth-century-style symmetrical architrave trim with turned corner blocks. The same type of trim is employed on the south front windows. Only the very small second-story windows on the east side of the east terminal wing retain both the original architrave trim with molded sills and the original wide-muntin eighteenth-century-style sash, probably because these windows open into an unfinished attic area completely inaccessible except by ladder from the exterior. The south entrance was also remodeled. A fanlight door with sidelights replaced what was probably a large, single-paneled door with a plain transom similar to the west entrance in the west terminal wing.

Chastellux described three porticoes at Battersea, and three porticoes can be seen when viewing the house from either the north or the south, one in the middle and one at each end. It is very difficult to determine if any of the porticoes existed in their present form at the time of Chastellux's visit. It can be said with certainty that the flat-roofed portico sheltering the south entrance did not exist in its current form in the eighteenth century. A break in the modillion cornice along the main roof indicates that the portico had two levels at one time. An examination of the roof framing shows no evidence that this upper level was ever connected to the main roof by a pediment, so it is assumed that the upper level had a flat roof. One can only guess whether or not this two-level portico was the original. Jefferson had a two-level portico on the first version of Monticello, but it was pedimented, fronting a temple-form center section. A two-level portico with a flat roof fronting a square center section topped by a pyramidal roof would be odd looking, to say the least, and not especially Palladian. It seems more reasonable that the south entrance was originaly treated with a dwarf portico similar in scale to the one that exists at the north entrance, and similar to that which was originally on Brandon. In any case, the present, rather wide portico, really a veranda, is probably not the bottom half of the two-level portico as it does not properly align with the breaks in the cornice.

The quite elegant one-bay Doric porticoes with their bold friezes of triglyphs and roundel metopes on each of the terminal wings are not likely original either, as they are much more Classical Revival in feeling than colonial. They probably date from the early nineteenth-century alterations and truly are a complement to the Palladian windows in the wings. Chastellux's description confirms that there were porticoes on the ends in the eighteenth century. Even the pedimented dwarf portico on the north entrance has undergone alterations and is of uncertain date. Despite all the changes in the windows, porticoes, and possibly stuccoing, the present state of Battersea presents an eminently appealing composition and evinces a conscious striving for upgrading and refining the house's appearance, a rare instance where alterations enhance rather than compromise the character a fine house was intended to have.

Battersea's interior changes are as extensive as those of the exterior. The chief original feature to have survived, and a really remarkable feature it is, is the matchless Chinese trellis stair in the narrow entrance hall, immediately in front of the saloon (fig. 7). The stair, the richest example of its type extant in Virginia, points, as in the case of Brandon, to the hand of Jefferson and his fondness for a Chinese accent. Though small in scale and cramped in its space, the stair comes across as very impressive and a quite grand approach to the upper floor, indicating a very talented designer. Jefferson's distaste for grand stairs is well known, but given the fact that the stair occupies a conservative amount of the first-floor space and does conform roughly to the plan in Morris's plate 3, a Jeffersonian influence should not be ruled out.

Except for the Chinese stair and the notably fine matching woodwork in the stair hall, the remaining original trim at Battersea is unremarkable, consisting mostly of paneled wainscoting and a paneled chimney wall in the main second-floor bedroom. Both the east hyphen room and the east terminal wing room were completely retrimmed in the early nineteenth-century remodeling with particularly handsome Federal woodwork, marble mantels, and plaster cornices. The saloon and dining room were largely retrimmed as well, but each retains its original paneled wainscoting.

These two houses, Brandon and Battersea, represent the epitome of Palladian influence in eighteenth-century Southside Virginia. Both, it has been shown, may have been designed by Virginia's preeminent statesman, philosopher, and architect, Thomas Jefferson. It is unfortunate in the extreme that the destruction of Jefferson's early papers will likely preclude the possibility of formally documenting the authorship of these two outstanding works by Palladio's chief advocate in the New World. However, the houses speak for themselves, clearly revealing the ultimate debt they owe for their graceful, timeless beauty to Italy's monumental architectural genius.

NOTES

¹ Fiske Kimball, *Thomas Jefferson, Architect* (Cambridge, Mass. 1916), p. 3.

² The "Frenchman's Map" of 1782, College of William and Mary, Swem Library; Mutual Assurance Society of Virginia, policy No. 592 for Littleton Tazewell, 1809, microfilm in Virginia State Library (reel 8, vol. 66, no. 972).

³ S. P. Moorehead, "Tazewell Hall: A Report on Its Eighteenth-Century Appearance," *Journal of the Society of Architectural Historians* 14, no. 1 (1944): 14–17.

⁴ Thomas Jefferson, *Notes on the State of Virginia*, ed. William Peden (Chapel Hill, N.C., 1955), p. 152.

⁵ Robert Morris, *Select Architecture*, 2d ed. (London, 1757), pl. 3.

⁶ Fiske Kimball, *Thomas Jefferson, Architect*, p. 34. Although Kimball states that Jefferson purchased *Select Architecture* in 1770–71, it is not unreasonable to assume that Jefferson owned an earlier copy of the book which was destroyed in the Shadwell fire. He claimed that virtually all of his books were lost in the fire.

⁷ John Winterbottom, indexer, *Genealogies of Virginia Families*, 3 vols. (Baltimore, 1981), 3:727, 756.

⁸ Thomas Tilleston Waterman, *The Mansions of Virginia, 1706–1776* (Chapel Hill, N.C., 1946), pp. 342–46.

⁹ Ibid., pp. 373–74.

¹⁰ Marquis de Chastellux, *Travels in North America in the Years 1780–81–82* (New York, 1828), p. 272.

1. *Villa Barbaro,* The Four Books of Andrea Palladio's Architecture, *ed. Isaac Ware (London, 1736), bk. 2, pl. 34.*

2. *Monticello, drawing by Thomas Jefferson, c. 1771. (Courtesy of the Massachusetts Historical Society)*

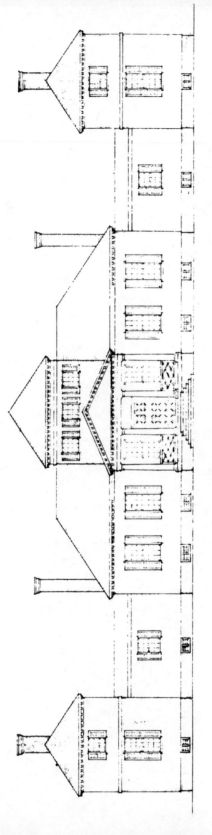

FRONT ELEVATION

▪ ▪ ▪ ▪ ▪

3. Tazewell Hall, reconstruction drawing by S. P. Moorehead. (Courtesy of the Colonial Williamsburg Foundation)

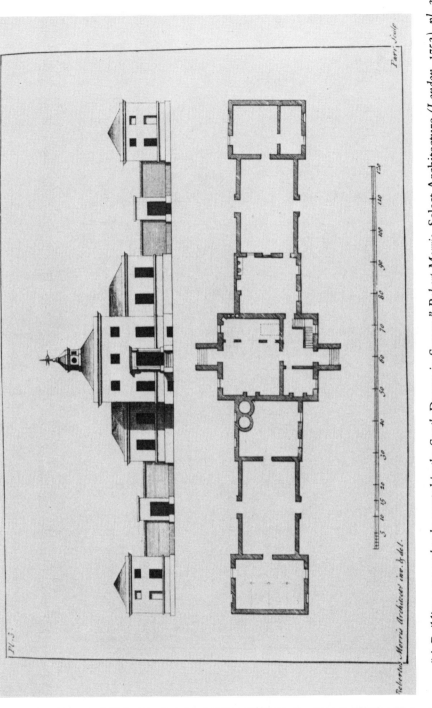

4. "A Building proposed to be erected in the South Downs in Sussex," Robert Morris, Select Architecture (London, 1752), pl. 3.

5. *Brandon, land (south) front. (Photograph by author)*

6. *Battersea, entrance (south) front. (Photograph by author)*

7. *Battersea, Chinese stair. (Photograph from the archives of the Virginia Historic Land-marks Commission)*

WILLIAM B. O'NEAL

PATTERN BOOKS IN AMERICAN ARCHITECTURE 1730–1930

It is necessary to begin this study with a very quick look at the English market for pattern books in the seventeenth and eighteenth centuries. Demand grew enormously between 1663 and the end of the eighteenth century. In 1663 the first book of Palladio together with Le Muet's essay on doors and windows was translated into English. The book's first French edition of 1645 is referred to by Professor Puppi as the "modest operation by Le Muet on the first book intended to organize a mere recipe for effective formulae for the use of the orders."

Godfrey Richards, the translator of that book, said in his introduction that there were "few books that we can recommend to you, beside the excellent Discourses of Sir Henry Wotton and John Evelin, Esq." The complete Palladio was not translated into English until 1715. But in 1724 William Halfpenny apologized for bringing out another pattern book "at a time when the Town is already Burthened with Volumes."

It was not until 1788, however, that I. and J. Taylor established a shop called the Architectural Library. Its list had fifty-nine entries at that time, but five years later the catalogue had grown to over a hundred items, including works by John Soane, James Paine, Hepplewhite, John Wood of Bath, Wil-

47

liam Chambers, Stuart and Revett, Brook Taylor, and Batty Langley.

Who bought these books? The subscription lists were filled with the names of gentlemen of title or with an "esquire" after their names. The next largest groups were clergymen, who were often also gentlemen, and, interestingly, physicians, who were not so frequently accepted as gentlemen. Architects were in a minority, while craftsmen usually appeared singly. Only the grander pattern books carried subscription lists, of course; the humbler handbooks were designed for use and low cost. They were often so well-used, as a matter of fact, as to be worn out, thus making them extremely rare on the present book market.

Next it is necessary to discuss the term *pattern book* since it is a phrase that is comparatively new, gaining its popularity since World War II.

Everyone knows what a book is—a gathering of leaves into a binding. It is an object which may range from the sumptuous illuminated manuscript within jeweled covers to the humblest of paperbacks.

A pattern, on the other hand, may vary from "the configuration of identically aimed rifle shots upon a target or the distribution and spread of shot from a shotgun" to "a dress pattern," or even "enough material to make a complete garment," and I am afraid, especially to the older generation, that a faint air of the seamstress's studio has permanently gathered about the word.

Nevertheless the first meaning of pattern is an archetype, that is, an original model or type after which other similar things are patterned. *Archetype* is a word ultimately derived from the Greek, meaning exemplary. *Pattern*, itself, is derived through Middle English and Middle French from the Medieval Latin *patronus*, which means not only "something to be imitated" but also patron, while *patron* in turn means both defender and advocate and derives from the Latin *pater* for father. All of which brings us back to the idea of the clone!

Are all architectural books pattern books? It is a question which may not always be answered affirmatively, and yet strange things happen to even the grandest of architectural books. There is no question, of course, about the many handbooks that have always been issued. In fact, the catalogue of the library of the Harvard School of Design still has a large section of books listed under handbooks. Their titles vary from such things as *The Country Builder's Assistant* and *The Young Builder's General Instructor* to *The City and Country Builder's and Workman's Treasury of Designs*.[1] The handbook tradition has continued until the present, but now the handbooks usually treat matters such as energy conservation, computer uses and applications, or long tables of figures about materials.

Then there was the sort of book which was a step above the handbook, but was obviously a pattern book. This was especially true of the books of Robert Morris in the mid-eighteenth century such as *Select Architecture* (fig. i), *Architecture Improved*, and *Rural Architecture*.[2] They were usually small quarto in size and issued with engravings superior to those in the majority of handbooks.

Morris even had a sense of humor, for in *Architecture Improved* he inserted this parody of an advertisement for a pattern book:

> There is now in Press, and speedily will be published, a treatise on Country Five-Barred Gates, Stiles, and Wickets, elegant Pig Sties, beautiful Hen Houses, and delightful Cow-cribs, superb Carthouses, magnificent Barn Doors, variegated Barn Racks and admirable Sheep Folds; according to the Turkish and Persian manner, a work never till now attempted. To which is added, some Designs of Fly Traps, Bee Palaces, and Emmet Houses in the Muscovite and Arabian Architecture; all adapted to the Latitude and Genius of England. The whole under the immediate Inspection of Don Gulielmus de Semi Je ne sçai Quoi, chief architect to the Grand Signor.[3]

It is not always so easy, however, to decide about a book. James Gibbs (1682–1754), for example, issued three books. His *Rules for Drawing the Several Parts of Architecture* was, of course, meant as a pattern book, for in it he states that "workmen once accustomed to his system will never follow any other." He included both St. Mary-le-Strand (fig. 2), his first public building in London after his return from Rome (where he had studied with Carlo Fontana, who himself had published the important *Templum Vaticanum* in 1695), and his later St. Martin's-in-the-Fields in his *Book of Architecture*.[4]

In the latter book he says that "Gentlemen . . . may here be furnished with Draughts of useful and convenient Buildings and proper Ornaments," so it, too, was obviously a copy book.[5] That it was so used may be easily proved by the number of Gibbs-like churches on the American East Coast. The book much increased Gibbs's influence and was purchased, aside from the gentlemen subscribers, by a carpenter, a carver, two doctors, six ecclesiastics, a judge, a merchant, six architects, one sculptor, and a painter.

With his third book we reach a more difficult classification. It is his *Bibliotheca Radcliviana*, an account of the building of the Radcliffe Library at Oxford.[6] It is clearly a self-congratulatory book in which he praises the trustees for whom he designed the building and lists and praises the principal workmen. I do not think he thought the building would serve as a pattern for future libraries; yet it was the first of many circular libraries, a scheme that was used right through the nineteenth century, both in England and in this country. So, willy-nilly, the book has also become a pattern book!

The most sumptuous architectural work is probably the great folio series of *The Antiquities of Athens*, lavishly bound in heavily gold-imprinted leather. The five volumes were an account, both literary and pictorial, of the findings of the first expedition into Greece by westerners in search of accurate information since its occupation by the Turks at the end of the Byzantine Empire. The small group reached Piraeus in

1751 and came back laden with measured drawings and pictorial views of the monuments in situ (fig. 3). The first volume did not appear until 1762, while the fifth was not issued until 1830. The best engravers in London, including William Blake, were engaged to execute the plates, which were often copied exactly by others, as in Peter Nicholson's *The Principles of Architecture* (fig. 4) or in some of the works of Asher Benjamin (fig. 5) in this country.[7] The conclusion reached is that even this very grand work is also a pattern book. In any case it was certainly conceived in order to bring accurate information about the buildings of the Greeks to Europe and America.

Thus it is apparent that almost any architectural book, no matter how humble or how splendid, falls into the category of a pattern book. This is probably also true of books of architectural history or theory or criticism since they, too, under certain circumstances may be used as patterns, in spite of the whiff of gunpowder and the aroma of the steam iron that have attached themselves to the term.

In this country there were four ways of gaining architectural knowledge before the Revolution. The pattern books were the principal literary method. One could be apprenticed to a builder. One could bring the memory of knowledge gained in the old country to the new. Or one could gain such knowledge through travel.

Pattern books were very much in use both by the builder-architect and by the gentleman-architect, who, if he was fortunate, combined travel with his books to complete his architectural education. Before the Revolution all such books were imported from Europe, usually from England, but, contrary to general opinion, the works of Palladio were not the most commonly held items in the builders' libraries up and down the coast, although his name was often invoked in books such as Salmon's *Palladio Londinensis*. This book was referred to no less than twenty-eight times by Thomas Tileston Waterman in his *Mansions of Virginia* in writing about the

Randolph-Peachy house, Marmion, Tuckahoe, Rosewell, Christ Church in Lancaster County, Sabine Hall, Nomini Hall, Westover, Shirley, Carter's Grove, and Wilton. James Pain's *British Palladio* was another of these to use Palladio's name, as was William Pain's *The Builder's Pocket Treasurer; or, Palladio Delineated.*[8]

In fact, Thomas Jefferson wrote James Oldham on December 24, 1804, that "there never was a Palladio here [in Washington] even in private hands till I brought one. . . . I send you my portable edition, which I value because it is portable. It contains only the 1st book on the orders which is the essential part."[9]

Nor were the architectural libraries of the time large. Jefferson had accumulated the largest, having forty-nine such works before he sold his "great" library to Congress in 1815. William Buckland, whose uncle in London, to whom he had been apprenticed, was both a joiner and an architectural bookseller, accumulated fifteen architectural works before his death in 1774. Of these at least six had been acquired after he had come to this country. William Byrd II of Westover had twenty-seven architectural entries in the catalogue of his library when it was offered for sale in 1777. The catalogue of the Carpenter's Company of Philadelphia lists only thirty-two such books printed before 1826. Samuel McIntire had seven architectural books at his death in 1811, while Charles Bulfinch could boast of only fifteen.[10]

With the end of the Revolution and the coming of the nineteenth century the deluge began, however. The first architectural book published in this country was a reprint (pirated?) of Abraham Swan's *The British Architect* in 1775.[11] A man named William Norman began to bring out a few, a very few, pattern books in the Boston area, while one of the William Pain handbooks was given a Philadelphia edition in 1797.[12]

But it was Asher Benjamin (1773–1845) who was the first American architectural author of some fame and extended productivity. He lived mostly in Greenfield, Massachusetts, and he designed many houses and churches in the area. His importance, though, springs from his books, through which he popularized, in turn, the late Georgian, then the Federal, and later the Greek Revival details. His books had no-nonsense titles such as *The Country Builder's Assistant*, *The Rudiments of Architecture*, and *The Builder's Guide*.[13]

They went through innumerable editions from the appearance of his first title in 1797 until the last in 1843. The illustrations were still engraved, and the volumes were usually sturdily bound in leather. It is no wonder they were carried west by the pioneers in their covered wagons, and it is still quite possible to find doorways, windows, or mantels that were copied directly from specific plates in Benjamin's books throughout the Midwest.

This same sort of book was also issued by Owen Biddle (first in 1805), by John Haviland (first in 1818), who went a step further and issued an "Improved" edition of Owen Biddle's *Young Carpenter's Assistant*, and by Minard Lafever, usually in smaller format and less sturdy binding than those used by Asher Benjamin. Of these neither Owen Biddle nor Minard Lafever seems to have gone beyond the handbook sort of pattern book, but John Haviland, who had become an expert in the design of penal architecture, began to answer the need for books on various sorts of buildings with his *Description of the New Penitentiary*, 1824, and his *Plans for the Halls of Justice*, 1835.[14]

The expansion of the nation to the west, the increased use of roads, the greater prosperity, and the larger population all led to the need for specialized as well as generalized pattern books. In addition, the increased ease of printing by mechanical means as well as the new ways of reproducing illustrations—the development of the wood line engraving, the steel engraving, and the lithograph—aided the publishers and the authors in satisfying this demand.

Ithiel Town as early as 1821 put out a book on the *Construction of Wooden and Iron Bridges*.[15] He not only published the

book, but he patented a bridge truss of such universal application that he became rich and one of the foremost architects
of the new nation, and he amassed the largest architectural
library known in the country until that time (there is one
estimate that it contained over ten thousand volumes, but
that surely must be an exaggerated figure). Town installed
the library in his house and opened it freely to anyone who
wished to study there.

A little later a man named Charles Ellet issued, within a
few years of each other, two books about wire suspension
bridges.[16]

Charles Bulfinch wrote two studies on prisons, one on
penitentiaries in 1827 and another on prisons of the Auburn
type in 1829. W. A. Alcott put out a book of schoolhouses in
1832, and Robert Mills began writing about lighthouses in
1837 and the Capitol of the United States in 1847–48. The
specialized pattern book, that is, one which deals with a single
building type, has been a feature of American publishing ever
since.[17]

Yet a new direction appeared with the opening of the door
for historical works. The first such work is an 1831 *History of
Sculpture, Painting, and Architecture* by J. S. Memes. The star of
the early histories, however, was William Dunlap's *Arts of
Design in the United States*, of 1834. Although Dunlap was not
a very good painter, he was vice-president of the National
Academy of Design. His book is not really a pattern book
since there are no illustrations, but he included architecture
in the "arts of design" and he went straight to the individuals
about whom he wrote for his information, or, if they were
no longer living, to their heirs or to existing documents. His
work has an invaluable amount of firsthand information about
this country's early artists and architects and has served well
many generations of scholars.[18]

If homes are added to the building types already listed, we
find the pattern of pattern books set for the rest of the century. The books were generally not as sumptuous as those of

the eighteenth century, at least those produced in this country were not, though it is possible to point to both English and French issues that were; those of Pugin (fig. 6) Hunt and Daly (fig. 7), for instance.[19]

Wood engraving and lithography were both used, especially the former, which became quite an expressive medium even in the very small vignettes that were frequently put into the texts. The books were now mostly cloth bound, frequently both blind stamped and gold stamped, and ranged from books that incorporated considerable theory to small pocket books whose purpose was only to give visual ideas.

Andrew Jackson Downing (1815–1852) was one of the best theorists of the mid-century. He had begun his training at his father's nursery but became, during his short life, well enough known in both architectural circles and in landscape architecture to be commissioned to design the grounds of the Capitol in Washington, the White House, and the Smithsonian Institution. He was also well enough known that his three books on homes—*Cottage Residences*, 1842, *Hints to Persons about Buildings*, 1859, and the *Architecture of Country Houses*, 1850, as well as that on *Landscape Gardening*, 1841—continued to be issued long after his death.[20] The fourth edition of *Cottage Residences*, for instance, followed the physical pattern of so many mid-century pattern books. It was bound in dark green pebbled cloth stamped in gold with a vignette of a Gothic Revival balcony surrounded with ivy. The spine has a design of a Gothic Revival chimney, also stamped in gold.

Calvert Vaux worked with Downing before he became associated with Frederick Law Olmsted in the design of Central Park in New York and went on to design the original Metropolitan Museum. Vaux went so far as to dedicate his *Villas and Cottages* of 1857 to Downing and his widow. Its engravings are even finer and richer than those in his former partner's books, and it is bound in brown pebbled cloth stamped in gold with a vignette of a garden gate.[21]

George Woodward, who described himself as an architect,

a civil engineer, and a publisher of architectural books, had a banner year in 1868 for he, himself, issued his *Country Homes* and Part II of his *Architecture and Rural Art*, Part I having been published in 1867. These are tiny volumes in duodecimo, obviously made to slip into a pocket and undoubtedly put out to appeal to a broad but not very moneyed public. They are bound in a serviceable dark garnet cloth, and the illustrations are steel engravings. Besides these works he also lists in the advertisement of his publishing house Daniel H. Jacques's *The House: A Pocket Manual* and Gervase Wheeler's *Rural Homes* and *Homes for the People*.[22]

One suspects that this format was very popular for there was an entirely unrelated pattern book put out as late as 1883 which was similar. It was called *House Plans for Everybody*, by S. B. Reed, who signed himself as an architect. It, too, is in dark garnet cloth, this time with a vignette of a small Victorian villa stamped in gold between black rules on the cover (fig. 8). Its engravings are less able than those of the earlier books, but the author included cost estimates for most of the buildings. One could, for example, have a half-stone, seven-room house with an inside bathroom for $2,800.[23]

Samuel Sloan is the last of these mid-nineteenth-century figures that we must examine. He not only practiced architecture up and down the East Coast from his base in Philadelphia, but he was a prolific writer who started, edited, and contributed to architectural magazines and who also issued a series of books from 1852 to 1873. They ranged from his first, *The Model Architect*, through books about city and country houses, churches, and proposals for the 1876 Centennial.[24] His books were of a rather larger format than the ones we have just been discussing, and he sometimes used the lithographic method for his illustrations. Through his literary activities he became even more prominent than through his architecture.

During the last quarter of the nineteenth century pattern books poured out of the presses in a steady and steadily increasing stream. There are many reasons for this. The greatest, of course, was the need for them, for if there had been no market they would not have been produced. But the causes of the need are various. After the Civil War the nation had turned from agriculture to manufacturing. This in time brought new wealth and widespread prosperity, neither of which was diminished as yet by the baleful income tax.

With the new wealth, large and larger architectural commissions could be awarded to the architects who were better and better trained both in the old methods and in the use of the developing new materials such as steel. A few young men were studying at the Ecole des Beaux-Arts in Paris, a few collegiate schools of architecture had been established, and the practice of architecture had been codified by the formation of the American Institute of Architects.

The combination of big business, steel, and a practical elevator made the skyscraper possible. With the big commission, whether it was for a commercial building, a factory, or the immense "cottages" of the Gilded Age, there also emerged the big architectural firm that advertised itself with a new kind of pattern book, the monograph on its own work. This was a natural development, since their clients often issued monographs about their own houses or their collections. The Vanderbilts and the Fricks both indulged in this hedonistic activity.

The development of the means to reproduce photographs vastly aided such schemes. The most notable of these monographs was the four-volume folio set of their own work sponsored by McKim, Mead and White. It was probably not thought of as a pattern book at all by its sponsors, and indeed their imitators never reached its heights, but the set appeared in many offices and libraries as well as in the hands of some prospective and prosperous clients.[25]

The tradition of the personal monograph has continued almost until today, but in very attenuated form due to the present costs of production. Before 1966 Philip Johnson said that he intended to sponsor a book about his work in the manner of the McKim, Mead and White opus, but when it was published it was a thin, spare, small folio about half the size of the others and only one instead of four volumes.[26] He was able, however, to illustrate his book entirely with color photographs, a device which was impossible for his predecessors.

In the twenties the photograph became the chief item in the pattern books while the text almost disappeared. It became the fashion to issue a book of photographs, often only of a particular era or area or both. The text would be restricted to an introduction and the necessary captions. Arthur Byne and Mildred Stapley were especially good at this; a single example of their books, *Spanish Interiors and Furniture*, gives the flavor, as does William Lawrence Bottomley's *Spanish Details*, which had only a single page of text.[27]

Fifty years ago students used to pore over these books, as well as the French photographic portfolios of the new architecture that was inspired by the Exposition des Arts Décoratifs of 1925—portfolios also of a single subject such as shop fronts or ironwork—in the hope that they could absorb the secret of the design of the objects pictured. That they never did did not discourage them.

But, like a thunderclap, a more serious "new" architecture came on the scene, a movement baptized the International Style by Henry-Russell Hitchcock and Philip Johnson for the 1932 exhibition at the Museum of Modern Art in New York. Their catalogue (fig. 9) itself became a pattern book, but it and the exhibition and the movement spawned a new kind of book for the architectural trade.[28] Solemn theory was suddenly the thing everyone was listening to, and the pattern books, while not giving up illustrations, were now primarily texts instead of photographs.

This appears even in the books of the giants of the twentieth century. In spite of the fact that Frank Lloyd Wright allowed his designs to be published in the *Ladies' Home Journal* early in the century and in entire issues of *House Beautiful* in November 1955 and October 1959 and supplied the pictorial material for the Wasmuth Press's two important publications, the *Ausgeführte Bauten und Entwürfe* of 1910 and the *Ausgeführte Bauten* of 1911, he also liked a preponderance of text to illustrations as in his *Autobiography* and his *Testament*.[29]

Walter Gropius, too, liked this method, finally minimizing each of the photographs in his *Scope of Total Architecture* to about the size of a postage stamp and reducing their number to the bare necessities.[30]

Le Corbusier, also, in spite of his lavishly illustrated multivolume *Oeuvre Complète*, indulged in the illustrationless work as well. His *When the Cathedrals Were White* is reduced to a few tiny, scratchy pen-and-ink sketches, while his *Talks with Students* omits them altogether. How impassioned his text was is easily seen in a passage from *When the Cathedrals Were White* in which he declared: "To examine their consciences and to repent, I challenge those who with the savagery of their hatred, fear, lethargy and indigence of spirit dedicate themselves with shameful violence to the undermining and destruction of whatever is most beautiful in this land of France and in this epoch."[31]

All of these were surely meant to be pattern books. At least, all of the three men were teachers and they all had enough self-respect to think themselves worthy of emulation.

But this discussion has been brought years beyond the assigned dates. Fifty years ago that influential exhibition of the International Style at the Museum of Modern Art was arranged by Johnson and Hitchcock. This spring Harvard held a seminar to celebrate it. It was, of course, both praised and condemned by the latter-day critics who were there. Philip Johnson, who was described by the *Washington Post* in its account of the affair as a "man who is a mean hand with a quip," was also there. In his summation he had the last word, for he pointed out that all he and Henry-Russell Hitchcock had tried to do when arranging the exhibition was to make an effect.

He added that they did and gave as his justification that the members of the seminar were, at that moment, still talking about it. It is good to see that pattern books have as much power as ever. Even the exhibition's 1932 catalogue is a very much respected reference today.

NOTES

[1] Asher Benjamin, *The Country Builder's Assistant: Containing a Collection of New Designs of Carpentry and Architecture Which Will Be Particularly Useful to Country Workmen in General* (Greenfield, Mass., 1797); Minard Lafever, *The Young Builder's General Instructor; Containing the Five Orders of Architecture* (Newark, N.J., 1829); and Batty Langley, *The City and Country Builder's and Workman's Treasury of Designs* (London, 1750).

[2] Robert Morris, *Select Architecture: Being Regular Designs of Plans and Elevations Well Suited to Both Town and Country* (London, 1755), *Architecture Improved, in a Collection of Modern, Elegant, and Useful Designs; from Slight and Graceful Recesses, Lodges, and Other Decorations in Parks, Gardens, Woods, or Forests, to the Portico, Bath, Observatory, and Interior Ornaments of Superb Buildings* (London, 1755), and *Rural Architecture: Consisting of Regular Designs of Plans and Elevations for Buildings in the Country, in Which the Purity and Simplicity of the Art of Designing Are Variously Exemplified. With Such Remarks and Explanations As Are Conducive to Render the Subject Agreeable* (London, 1750).

[3] Morris, *Architecture Improved*, p. ii.

[4] James Gibbs, *Rules for Drawing the Several Parts of Architecture in a More Exact and Easy Manner Than Has Been Heretofore Practised* (London, 1732), pp. 1–2, and *A Book of Architecture Containing Designs of Buildings and Ornaments* (London, 1728).

[5] Gibbs, *A Book of Architecture*, p. i.

[6] James Gibbs, *Bibliotheca Radcliviana; or, A Short Description of the Radcliffe Library, at Oxford* (London, 1747).

[7] James Stuart and Nicholas Revett, *The Antiquities of Athens*, 5 vols. (London, 1762–1830); Peter Nicholson, *The Principles of Architecture, Containing the Fundamental Rules of the Art, in Geometry, Arithmetic, & Mensuration* (London, 1809), pl. 91; and Asher Benjamin, *The Architect or Practical*

House Carpenter: Being a Complete Development of the Grecian Orders of Architecture (Boston, 1839), pl. 4.

[8] William Salmon, *Palladio Londinensis; or, The London Art of Building* (London, 1748); Thomas Tileston Waterman, *The Mansions of Virginia, 1706–1776* (Chapel Hill, N.C., 1946); James Pain, *The British Palladio; or, The Builder's General Assistant* (London, 1788); and William Pain, *The Builder's Pocket Treasurer; or, Palladio Delineated and Explained* (London, 1785).

[9] Millicent Sowerby, *Catalogue of the Library of Thomas Jefferson*, 5 vols. (Washington, D.C., 1952–59).

[10] Fiske Kimball, *Thomas Jefferson, Architect* (Boston, 1916), pp. 90–100; *Buckland: Master Builder of the 18th Century* (Lorton, Va., 1978), pp. 13–16; advertisement, *Virginia Gazette*, Dec. 19, 1777; Carpenter's Company of the City and County of Philadelphia, *Finding List of the Library* (Philadelphia, 1894); Fiske Kimball, *Mr. Samuel McIntire, Carver, Architect of Salem* (Portland, Maine, 1940), p. 23; and Charles A. Place, *Charles Bulfinch, Architect and Citizen* (Boston, 1925), pp. 285–88.

[11] Abraham Swan, *The British Architect: A Collection of Designs in Architecture* (London, 1745); Swan, *The British Architect* (Philadelphia, 1775).

[12] William H. Jordy, comp., *Chronological Short Title List of Henry-Russell Hitchcock's "American Architectural Books"* (Charlottesville, Va., 1955), p. 1.

[13] The works of Asher Benjamin have not yet been thoroughly explored, but there are no less than seven titles: *The Country Builder's Assistant* (1797), *The American Builder's Companion* (1806), *The Rudiments of Architecture* (1814), *The Practical House Carpenter* (1830), *Practice of Architecture* (1833), *The Builder's Guide* (1839), and *Elements of Architecture* (1843), and many editions of each given in Jordy, pp. 1–3.

[14] Owen Biddle, *The Young Carpenter's Assistant: or, A System of Architecture, Adapted to the Style of Buildings in the United States* (Philadelphia, 1805); John Haviland, *The Builder's Assistant, Containing the Five Orders of Architecture, Selected from the Best Specimens of the Greek and Roman, with the Figured Dimensions of Their Height, Projection, and Profile, and a Variety of Mouldings, Modillions, & Foliage on a Larger Scale, Both Enriched and Plain; with Working Drawings Showing Their Method of Construction; Selected from a Number of Beautiful Examples, Copied from the Antique. For the Use of Builders, Carpenters, Masons, Plasterers, Cabinet Makers, and Carvers* (Philadelphia, 1818–21), and *An Improved and Enlarged Edition of Biddle's Young Carpenter's Assistant; Being a Complete System of Architecture for Carpenters, Joiners, and Workmen in General, Adapted to the Style of Building in the United States. Revised and Corrected, with Several Additional Articles and Forty-eight New Designs, Chiefly of Full Size Working Drawings of Modern Finish, in*

Detail (Philadelphia, 1837); Minard Lafever, *The Modern Builder's Guide. Illustrated with Eighty-eight Copper-plate Engravings* (New York, 1833), *The Beauties of Modern Architecture, Illustrated by Forty-eight Original Plates Designed Expressly for This Work* (New York, 1835), and *The Modern Practice of Staircase and Handrail Construction, Practically Explained in a Series of Designs, with Plans and Elevations for Ornamental Villas* (New York, 1838); John Haviland, *A Description of Haviland's Design for the New Penitentiary, Now Erecting near Philadelphia. Accompanied with a Bird's Eye View* (Philadelphia, 1824), and *Plans for the Halls of Justice, New York* (New York, 1835).

[15] Ithiel Town, *Improvement in the Construction of Wooden and Iron Bridges Intended as a General System of Bridge Building for Rivers, Creeks, and Harbors, of Whatever Kind of Bottoms, and for Any Practicable Width of Span or Opening, in Every Part of the Country* (New York, 1821).

[16] Charles Ellet, *Report and Plan for a Wire Suspension Bridge Proposed to Be Constructed across the Mississippi River at Saint Louis* (Philadelphia, 1840), and *A Popular Notice of Suspension Bridges, with a Brief Description of the Wire Bridge across the Schuykill, at Fairmont* (Philadelphia, 1843).

[17] Charles Bulfinch, *Bulfinch on Penitentaries. Report of Charles Bulfinch on the Subject of Penitentiaries* (Washington, D.C., 1827), and *Statement of Charles Bulfinch on the Construction and the Physical and Moral Effects of Penitentiary Prisons of the Auburn Type* (Washington, D.C., 1829); William Andrews Alcott, *Essay on the Construction of School Houses* (Boston, 1832); Robert Mills, *The American Pharos, or Light-house Guide* (Washington, D.C., 1832), and *Guide to the Capitol and National Executive Offices of the United States* (Washington, D.C., 1847).

[18] John Smythe Memes, *History of Sculpture, Painting, and Architecture* (Boston, 1831); William Dunlap, *History of the Rise and Progress of the Arts of Design in the United States* (New York, 1834).

[19] A. W. N. Pugin, *Specimens of Gothic Architecture; Selected from Various Ancient Edifices in England* (London, 1821), and *The True Principles of Pointed or Christian Architecture* (London, 1841); T. F. Hunt, *Exemplars of Tudor Architecture, Adapted to Modern Habitations* (London, 1836); César Daly, *L'architecture privée aux XIXme siècle sous Napoleon III*, 2 vols. (Paris, 1864).

[20] Andrew Jackson Downing, *Cottage Residences; or, A Series of Designs for Rural Cottages and Their Gardens and Grounds. Adapted to North America* (New York, 1842), *Hints to Persons about Building in the Country* (New York, 1859), *The Architecture of Country Houses; Including Designs for Cottages, Farm Houses, and Villas, with Remarks on Interiors, Furniture, and the Best Modes of Warming and Ventilating* (Philadelphia, 1850), and *A Treatise on the Theory*

and Practice of Landscape Gardening, Adapted to North America; with a View to the Improvement of Country Residences (New York, 1841).

21 Calvert Vaux, *Villas and Cottages. A Series of Designs Prepared for Execution in the United States* (New York, 1857).

22 George E. Woodward, *Woodward's Country Homes* (New York, 1868), *Woodward's Architecture and Rural Art*, no. 1 (New York, 1867), and *Woodward's Architecture and Rural Art*, no. 2 (New York, 1868); Daniel Harrison Jacques, *The House: A Pocket Manual of Rural Architecture: Or How to Build Country Houses and Outhouses* (New York, 1866); Gervase Wheeler, *Rural Homes: or, Sketches of Houses Suited to American Country Life, with Original Plans, Designs, &c* (New York, 1865), and *Homes for the People, in Suburb and Country; the Villa, the Mansion, and the Cottage, Adapted to American Climate and Wants* (New York, 1864).

23 S. B. Reed, *House Plans for Everybody: For Village and Country Residences, Costing from $250 to $8,000; Including Full Descriptions and Estimates in Detail of Materials, Labor, and Cost, with Many Practical Suggestions and 175 Illustrations*, 5th ed. (New York, 1883).

24 Samuel Sloan, *The Model Architect. A Series of Original Designs for Cottages, Villas, Suburban Residences, etc., Accompanied by Explanations, Specifications, Estimates, and Elaborate Details* (Philadelphia, 1852), *American Houses: A Variety of Original Designs for Rural Dwellings* (Philadelphia, 1861), *City and Suburban Architecture: Containing Numerous Designs and Details for Public Edifices, Private Residences, and Mercantile Buildings* (Philadelphia, 1859), *City Homes, Country House, and Church Architecture* (Philadelphia, 1871), and *Description of Design and Drawings for the Proposed Centennial Buildings, to Be Erected in Fairmount Park* (Philadelphia, 1873).

25 *A Monograph of the Work of McKim, Mead & White, 1879–1915*, 4 vols. (New York, 1915–17).

26 Henry-Russell Hitchcock, intro., *Philip Johnson: Architecture, 1949–1965* (New York, 1966).

27 Arthur Byne and Mildred Stapley, *Spanish Interiors and Furniture* (New York, 1928); William Lawrence Bottomley, *Spanish Details* (New York, 1921).

28 *Modern Architecture: International Exhibition* (New York, 1932).

29 Frank Lloyd Wright, *Ausgeführte Bauten und Entwürfe von Frank Lloyd Wright* (Berlin, 1910), *Ausgeführte Bauten* (Berlin, 1911), *An Autobiography: Frank Lloyd Wright* (New York, 1932), and *A Testament* (New York, 1957).

30 Walter Gropius, *Scope of Total Architecture* (New York, 1955).

[31] Le Corbusier, *Oeuvre Complète*, 7 vols. (Zurich, 1946–66), *When the Cathedrals Were White: A Journey to the Country of Timid People* (1947; rpt. New York, 1964), p. 3 and *Le Corbusier Talks with Students from the Schools of Architecture* (New York, 1961).

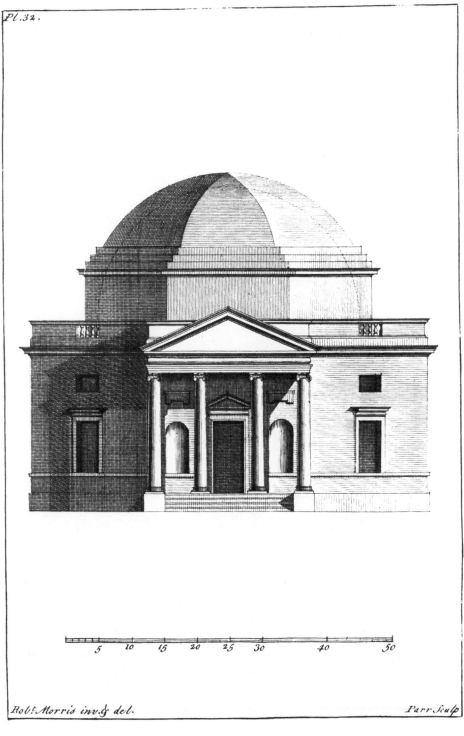

Robt Morris inv & del.

Parr Sculp

1. *Elevation, an octagonal temple or chapel, Robert Morris,* Select Architecture *(London, 1757), pl. 32.*

Prospectus Templi Sᵗᵃ Mariæ Londini in vico dicto the Strand, architectura Iacobi Gibbs.

Ja: Gibbs Archi: inven: et del: Jo: Harris fs.

2. *St. Mary-le-Strand, James Gibbs,* A Book of Architecture *(London, 1728), pl. 21.*

3. "*A view of the Tower of the Winds in Its Present Condition,*" *James Stuart and Nicholas Revett,* The Antiquities of Athens *(London, 1762), vol. 1, pl. 1.*

GRECIAN ARCHITECTURE

From the Temple of Theseus at Athens.

Drawn by P. Nicholson.

Engraved by W. Lowry

London, Published Sept. 1796, by P. Nicholson & C.º

4.   "Grecian Architecture: From the Temple of Theseus at Athens," Peter Nicholson, Principles of Architecture *(London, 1809), vol. 3, pl. 128.*

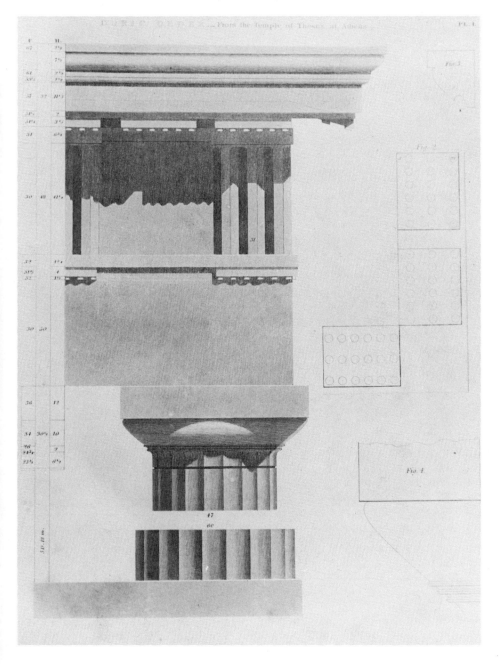

5. *Asher Benjamin,* The Architect or Practical House Carpenter *(Boston, 1839),*
pl. 4.

Specimens

of

Gothic Architecture:

selected from various

Ancient Edifices

in

England:

by

A. Pugin, Arch.t

Henry the 7.th Chapel, Entrance to N. aisle.

6. Title page, A. W. N. Pugin, Specimens of Gothic Architecture *(London, 1821).*

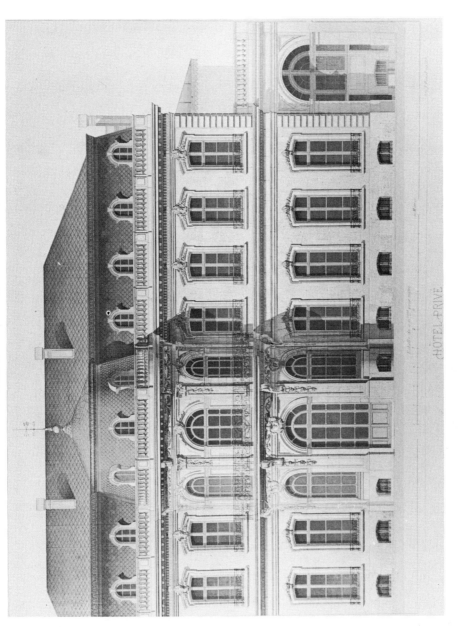

HÔTEL PRIVÉ

7. *Hôtel privé, César Daly, L'architecture privée aux XIXme siècle sous Napoleon III (Paris, 1864), vol. 1, pl. 1.*

8.   *S. B. Reed,* House Plans for Everybody *(New York, 1883).*

MODERN ARCHITECTURE

MUSEUM OF MODERN ART

9.   *Cover,* Modern Architecture: International Exhibition *(New York, 1932)*.

WILLIAM M. S. RASMUSSEN

PALLADIO IN TIDEWATER VIRGINIA: MOUNT AIRY AND BLANDFIELD

In the 1790s the English traveler Isaac Weld journeyed through northern tidewater Virginia, where he met gentlemen he described as well educated and living "in a style, which approaches nearer to that of English country gentlemen, than what is to be met with any where else on the continent."[1] He viewed immense estates in that region that had been established during the colonial era by such dynasties as the Tayloes of Mount Airy (figs. 1 and 2) and the Beverleys of Blandfield (see figs. 7 and 8). Close neighbors across the Rappahannock River, the Taylors and Beverleys were among the most prominent of colonial families, leaders in both government and agriculture and exemplars of an English standard of living not forgotten in the New World.[2] Much like the "noble Venetians" from whom Palladio designed the country houses depicted in his *Four Books of Architecture*, these noble Virginians developed an affinity for Renaissance architecture.

Virginians had long been accustomed to the utilitarian and utopian situations Palladio described as sites for the country retreats of his Italian contemporaries. He advocated the placement of villas in rural settings akin to those found by the earliest settlers in tidewater Virginia, where tobacco-producing plantations from the start were located whenever possible on navigable rivers: "If one may build upon a river, it will be both convenient and beautiful; because at all times and with little expence, the products may be convey'd to the city in boats, and will serve for the uses of the house and cattle. Besides cooling the air in the summer very much, it will afford a beautiful prospect."[3] By the mid-eighteenth century, when the masters of Mount Airy and Blandfield gathered inherited funds to build mansions befitting their status, Palladian ideas revived a generation earlier in England by Lord Burlington made their way to the American colonies. They were quickly and easily adapted in Virginia. Thus Mount Airy, begun close to 1754, about fifteen years before Blandfield, is distinguished as the first house in America to achieve Palladio's plan for an expansive five-part "villa."[4] If not quite as early, Blandfield owes its grandeur to the same Palladian scheme.

The first of the Tayloe family in Virginia arrived from London over a hundred years before the present Mount Airy was built. William Tayloe bequeathed his property to a nephew of the same name who in 1683 is said to have built the "Old House" at Mount Airy, located a few miles southwest of its successor and closer to the river. This William Tayloe married Anne Corbin, of the family that would build nearby Peckatone.[5]

Their son, John Tayloe I, who lived in his father's "Old House," served in the House of Burgesses and on the Governor's Council and became one of the wealthiest men in Virginia. When Tayloe died in 1747, his property was valued at over £10,000, an enormous estate comparable to that of his very successful contemporary in Virginia, Robert ("King") Carter. Although the "Old House" does not survive, its inventory indicates the cultivated manner of living there.[6] Perhaps the father's structure had been enlarged by the son, for in 1747 the house must have been of similar size to its successor: on the ground floor, in addition to a "Passage" and a "Dining Room," were three richly furnished rooms, a "Hall," a "Green Room" (a bedchamber), and Mrs. Tayloe's "Chamber" (another bedchamber); the second level contained the areas described as "the Room above Mrs. Tayloe's," the "Great Chamber," another "Chamber," and "the Room over the green Room." As there were other rooms too, including the "Back Passage" and the "Inner Room," the "Old House" probably lacked the balance and symmetry offered in Palladian designs. So a new house was built, undoubtedly well furnished with many of the expensive tables, pieces of china, and bedsteads inventoried in the "Old House." With his inheritance John Tayloe II erected a great Palladian house where he became "renowned for his hospitality," tending to such genteel pursuits as even the instruction of servants in music "for the entertainment of his guests."[7]

Mount Airy was presumably named, as Edith Sale deduced in 1927, "because the beautiful spot seemed so high up among the clouds."[8] An article by Arthur Brooke published in 1899 suggests that the house was designed by a Colonel Thornton, "a friend of Colonel Tayloe, stationed with the army in Virginia, . . . perhaps kinsman of that Dr. William Thornton, physician, whose name was associated with the erection of the Capital at Washington."[9] William Thornton designed the Octagon House (1798–1800), John Tayloe III's Washington winter residence, so perhaps Brooke simply confused his Thorntons and Tayloes. A letter of 1754 from Edmund Jenings of Maryland suggests that John Tayloe II was more than likely the designer of his new house. According to his will the first John Tayloe had turned to Jenings "for his friendly advice upon several occasions."[10] The younger Tayloe sought similar guidance from his father's friend, who responded to Tayloe's "Intention of building on the Hill." Advising that "Every Scituation . . . must be adapted to y[ou]r Country & Climate," Jenings referred to the "Cold . . . & Large Familys & frequently much Compa[ny]" in Virginia and suggested "what I recommended to Col. Corbin—w[i]th Porticoes before the wings."[11] As a portico is defined as a covered ambulatory, this apparently is a reference to the connecting hyphens of Palladio's country villas. Hyphens were appropriate because they offered shelter from the "Cold" mentioned by Jenings and because linkage to the main block made the dependencies more functional, almost doubling the accommodations for "Large Families" and "much Company." From our limited knowledge of Peckatone, which does not survive, Jenings's advice was well heeded by Tayloe but ignored by Corbin. Jenings left the "Dimensions" of the "whole or any part" to Tayloe's "Discretion" but added that "Closets on y[ou]r North or NE part of y[ou]r House are Generally Damp without Continual Fires very near them." As we will see, this was one of several factors that apparently caused Tayloe to place his stairs in the supposedly "damp" northeast corner of the house (fig. 3). He was building in 1759, for a notation in his account book of that year refers to a payment of £200 "for my house."[12]

Although Palladio was translated into English in the eighteenth century, for architectural advice gentlemen amateurs in the colonies turned frequently by mid-century to simpler and more practical publications like James Gibbs's 1728 *Book of Architecture*, one of many how-to-do-it products of the Age of Reason. Its author stated that this book, which seems to have influenced Tayloe in several respects, was especially valuable "in the remote parts of the Country, where little or no assistance for Designs can be procured."[13] Architects like Gibbs had, of course, read Palladio. They saw an impressive and practical invention in his five-part plans, which Palladio very sensibly explained would allow the master "to go every place under cover, . . . neither the rains nor the scorching sun of the summer [would] be a nuisance to him."[14] From these plans and from Palladio's motif of three central arches (fig. 4), Gibbs developed the five-part Palladian design copied on the river facade at Mount Airy with unusual fidelity (fig. 5). The distinguishing motif of Palladio's Basilica at Vicenza was transformed by Georgian architects into a "Palladian window" that Gibbs thought contributed an "uncommon, but . . . good effect"; here was precedent for Tayloe's southeast Palladian windows. Tayloe's placement of chimneys within the building was exceptional in Virginia's warm summer climate; his stairs seem to have been situated in a corner of the house, a location unusual in Virginia designs; one room, the main parlor, is larger than the others; and the hall is recessed on both major facades, sheltered against the elements in the protective manner of Palladio. Within the recesses, even the niches at Mount Airy find precedent in Gibbs (plate 108).[15]

As most Virginia Georgian mansions were built of brick, Mount Airy is conspicuous in its use of dark local stone trimmed with light freestone from the Aquia quarries upriver on the Rappahannock.[16] The evidence seems contradictory as to whether or not Mount Airy once looked different. If we can believe a lithograph (fig. 6) published by Pendleton and Company before a severe fire in 1844, it was once stuccoed and painted white. Very obvious evidence of old stucco and white paint still remains on two outer dependencies. The stone of the large dependencies appears to have been gouged in preparation for stucco, and the windows project sufficiently to allow for the thickness of a hard covering. But if stucco and/or paint were applied, why does no trace remain today, particularly on the large dependencies, which were not at all damaged by the fire?[17]

Whether or not the center house and its connected outbuildings were actually stuccoed, such a plan certainly was intended at one point. Although it was fashionable in the 1820s to paint a building white to resemble the stone structures of antiquity, colonial examples of this practice are not unknown. Menokin, the adjoining estate given by the builder of Mount Airy to his daughter Rebecca on her marriage to Francis Lightfoot Lee in 1769, though now in ruins, has retained its stucco. Its plan and elevation, which have somehow survived, seem to indicate that stucco and white paint were intended from the start. The light stone trim of Mount Airy is absent; without stucco the dark quoins and belt courses of Menokin would have been barely visible. Neighboring Nomini

Hall, according to the tutor at that house in 1774, was "perfectly white."[18] Mount Pleasant, a Philadelphia house constructed in 1761, was stuccoed in the manner intended for Mount Airy. Not the least consideration regarding the use of stucco at any of these houses is the suggestion by James Gibbs in several instances in his *Book of Architecture* that "the Portico, Windows, Fascia's, Entablature, and all the projecting parts . . . be of Stone, and the rest of Brick finish'd over with Stucco."[19] Of course, Gibbs was only following the example of Palladio, who constructed many of his buildings of cheap material covered with stucco in imitation of stone.

The 1844 fire reportedly was started by a deranged slave while the owner was in Alabama tending to cotton operations and the family was at church. Furniture and paintings were saved, but in the words of a witness for William A. Tayloe's insurance claim, of "the centre building of Mount Airy house, all except the walls [was] consumed by fire."[20] The building must have then looked like Mannsfield, the Palladian house of Tayloe's sister, after it was burned in the Civil War. In 1846 Tayloe engaged Washington joiners George H. and William P. Van Ness to rebuild Mount Airy. They agreed to refinish the interior and repair the stonework injured by the fire.[21]

In the Pendleton lithograph Mount Airy is rendered totally white.[22] In the pediment is an elaborate crest, above a recessed entrance with three arched bays, all different from what is found today. In 1774 the visiting tutor Philip Fithian placed the "Lodging Rooms," including the nursery, in the large southeastern dependency.[23] An 1805 insurance policy indicates the structure was an "office used as part of dwelling," and in 1899 it was identified by Brooke as "the living wing," where the family could find privacy in low-ceiling rooms, each with its own ample fireplace. After decades of disuse this building's interior is being restored to its original appearance. To the left of this "living wing," a thirty-six-foot cube, is an eighteen-foot stucco structure. Included in the formal plan and linked by a fence with urns on pedestals as at the center, it was described in the 1805 insurance policy as a stone schoolroom.[24] Its doorway was designed after a scheme by Palladio as published in another of the English architectural handbooks.[25]

Insurance policies taken through the Mutual Assurance Society of Richmond in 1797, 1805, and 1816 place a number of structures west of the central buildings. These include a stone dairy, two stone coach houses (destroyed), a wooden smokehouse, a brick counting room (an office), a brick stable, and the spectacular brick orangery or greenhouse that is part of the garden plan.[26]

The exterior of the main house at Mount Airy is typical of the mid-Georgian period in which it was built in its use of a central pavilion and pediment to achieve the emphasis at the center advocated by Palladio. Both the plan and elevation to the chimney tops are close in dimension to a square-and-a-half, a simple shape, Palladio explained, that seems "correct" to our eyes. As he might place double giant pilasters at the corners for termination, the effect is achieved at Mount Airy by quoins. The original cornice and roof replaced after the fire probably resembled the modillioned cornice and double-hip roof still at Blandfield. At the side elevations, where the designer got very little help from architectural books, Palladian attention to the center and ends is repeated. An unusual effect is created by a doubling of the belt course.

The interior of Mount Airy radiates around its great lower hall, totally free of stairs in the manner almost always followed by Palladio. As described by Isaac Weld in the 1790s, the hall in a Virginia house "is always a favourite apartment, during the hot weather, . . . and is usually furnished similar to a parlour, with sofas, &c."[27] This hall would have served for dining as well as sitting during warm weather, after which it must have been used less frequently since no fireplaces open into it. But a lack of physical heat apparently proved no hindrance for large New Year's festivals attended here by sixty persons. These took place one day at Mount Airy, followed the next at neighboring Sabine Hall, whose master Landon Carter recorded them in his diary.[28]

Before the fire, tradition tells of Mount Airy interiors finished with marble floors, mahogany wainscoting mounted in silver, and a French cut-glass chandelier which John Tayloe III refused to sell to the White House after that building was destroyed in 1814.[29] Whether such accounts are entirely or partially true, Mount Airy was indeed spectacular in the eighteenth century for another reason. Its wainscoting, which would have been only chair-rail height in this mid-Georgian period, was carved by William Buckland. George Mason brought the well-trained and talented London joiner to Virginia to finish Gunston Hall. Buckland later practiced in Annapolis, where the Tayloes had many family relatives and friends. We have to envision interiors at Mount Airy of similar magnificence as Buckland left elsewhere. He purchased a farm very near to Mount Airy and Sabine Hall in 1767, and his name appears in the Mount Airy accounts between 1761 and 1770, particularly during the period of 1761–64.[30] Fragments of cornice that survived the fire were converted into mantel shelves and give us the only physical evidence of Buckland's presence at Mount Airy, which must have thus been unsurpassed in magnificence in all the English colonies. As Buckland also sold chairs, perhaps his hand is evident in some furniture at houses such as Mount Airy.[31]

Placement of the stairs directly southeast of the hall in John Tayloe's house would have been difficult, for there are two Palladian windows at mid-level on the southeast facade, one on each floor. Today, that on the ground level is a true window only in the center section, the one portion visible from within the house.[32] Furthermore, the southern interior wall to the southeast is solid masonry, as is the one opposite the stairs and now very visible in the ruins of the large Rosewell hall. As mentioned earlier, placement of stairs in a corner room was not unknown to Gibbs (or to other authors), and the space to the northeast was evidently considered undesirable because of potential dampness. Most important, Arthur Brooke, in 1899 on information given him by Henry Augustine Tayloe, who knew the prefire house as a young boy, placed the original stairs in the northeast corner, as did F. C. Baldwin in an article in 1913.[33] To reach the stairs built by the Van Nesses after the fire, one must pass through the modern dining room.

The tutor Philip Fithian also called this space a "Dining Room" when he visited just before the Revolution, although "parlor" might have been a more proper term since this is the largest room after the hall and thus would have served as the principal area for sitting as well as dining during the cold months.[34] Here, according to Fithian, "besides many other fine Pieces, are twenty four of the most celebrated among the English Race-Horses, Drawn masterly, & set in elegant gilt Frames." Many of the "Forty-seven mahogany chairs" valued at $295 for the 1815 property tax would have been used in the hall and in here, where "Three mahogany Dining Tables" valued at $75 were undoubtedly kept in winter.[35] Research indicates that some of these pieces were made in colonial Williamsburg.[36]

Adjoining this large dining room or parlor is the smaller space that since 1846 has housed the stairs. This space may initially have served as a "small dining room," in the manner of those documented at such Virginia houses as Sabine Hall and Nomini Hall.[37] This would have been a dining place for the likes of children and their attendants (the builder of Mount Airy had nine children), who might be best served separate from important guests. Or, isolated between the kitchen and large dining room, it could have simply served for storage of china, glass, and silver. After the fire of 1844 this room apparently no longer functioned as it once had. Thus the stairs were placed here, freeing the opposite northeast corner to be a new room, once a twentieth-century parlor, now used as a bedchamber.

One plate given by James Gibbs (plate 55) that is similar to the plan of Mount Airy places the stairs and library to one side of the center passage. Possibly the southeast lower room in Tayloe's house, adjacent to the original stairs and now a library, initially functioned as a library. For somewhere—probably on this floor—was the "excellent library" to which tutor Samuel Ripley noted in 1804 that he had "free access."[38]

The upper floor at Mount Airy features a cross hall that runs the length of the building, allowing three rooms to the south, with two rooms and the stairs to the north. Except for the reversed stairs, this basic plan, which is found at Bland-field and which allows a maximum number of rooms, prob-ably precedes the fire. This is suggested by a description in 1827 by an English visitor, Anthony St. John Baker: "Large hall, in centre, through the house; upstairs, a long gallery, with family portraits: the Corbins, Platers, &c."[39] Apparently the ground-floor plan was not repeated above, where hung the handsome family portraits (still at Mount Airy) of John Tayloe II and his wife Rebecca Plater, her brother Governor George Plater of Maryland (all by John Wollaston), and per-haps the portraits of Governor Samuel Ogle of Maryland and his wife, whose granddaughter married John Tayloe III. In English fashion this "long gallery" provided not only passage to the upper rooms but also an area for the exercise of walk-ing, which was undoubtedly enhanced by contemplation of the images of such accomplished family members.

The upper rooms off this "long gallery" must have been those Arthur Brooke described in 1899 as "guest chambers . . . on as grandiose scale [as the hall]."[40] Most of these seem to have been reserved for the frequent guests entertained here.[41] They would have been well furnished with many of the "Eight Mahogany Bedsteads," fine "looking Glasses," and several chests of drawers for which John Tayloe III was heavily taxed in 1815.[42] His total payment, $190.30¼, was twice the amount at either Blandfield or Sabine Hall, an indication of the mag-nificent standard of living maintained at Mount Airy for many generations.

The English visitor of 1827 also mentioned the front ter-
raced lawn at Mount Airy "with flowers on pedestals," de-
scending to a deer park, where family and guests took long
walks.[43] On occasion they must have experienced difficulties
with the "many fine deer" noted there by Baker, for George
Washington's gardener at Mount Vernon was often bitten by
his deer, which also ate shrubs. "I am at a loss," wrote the
general, "in determining whether to give up the Shrubs or
the Deer."[44]

The formal gardens to the south of the house, which Baker
described as "extensive," were no doubt designed by builder
John Tayloe II with the aid of his copy of Philip Miller's popular *Gardener's Dictionary*, which he purchased in 1760.[45] Arthur Brooke sketched in 1899 how the garden must have looked
in the colonial period, noting that "paths and ornaments" in
his day could "still be traced." His rendering matches the description in 1774 by the tutor Fithian as "a large well formed,
beautiful Garden, as fine in every Respect as any I have seen
in Virginia." Fithian added that "in it stand four large beautiful Marble Statues."[46] Brooke mentioned the spacious bowling
green in the center, flanked by the formal parterre to the east
and boxwood to the west that seemed "mountainous." Some
of the more sightly vegetables probably were planted here.
Beyond the boxwood remain the ruins of a spectacular orangery like the one in Maryland at Wye House owned by the
Lloyds, into which family a daughter of John Tayloe II married in 1767. According to the early nineteenth-century insurance policies, its center section was a "Green house" covered
with a wooden roof, flanked on each end by a "Hot house"
with "sides of glass" and "cov[ere]d with glass." In 1827 note
was made of "orange and lemon trees put out on the grass"
before this structure, where raspberries were produced at about
this same time and sent to Yorktown for an October dinner
in honor of the visiting Lafayette.[47] Below the bowling green
more vegetables would have been planted, flanking a broad
path bordered with fig trees on axis toward the river.[48]

Across the Rappahannock River and only a few miles upstream from Mount Airy is Blandfield (figs. 7 and 8), a house that with the passage of two centuries has not been remembered like its neighbor. Until modern times the Beverley home was somewhat isolated from activities to the north, due to the difficulties of travel about which Isaac Weld complained in the 1790s: "It is a most irksome piece of business to cross the ferries in Virginia; there is not one in six where the boats are good and well manned, and it is necessary to employ great circumspection in order to guard against accidents, which are but too common."[49]

At Mount Airy the gentle curves of the connecting quadrants soften the rigid lines they unite. But as well as curved ones, Palladio also recommended angular connecting quadrants like those built at Blandfield with attention to proportion. If the wide 140-foot total front of this "country villa" seems "correct" as well as impressive, it is because that dimension is twice the seventy-foot front of the house itself. It in turn is eight feet deeper than Mount Airy, out of Palladian proportion in this respect to be grander in scale.

The Beverley family was distinguished when in 1705 the second Robert Beverley in Virginia, known as the historian, published in London his *History and Present State of Virginia*, a well-written book particularly valuable for its early information. Born in Middlesex County, Virginia, and educated in England, he served in public positions in the colony as clerk and burgess, while cultivating on his plantations huge vineyards and accumulating for his son one of the largest landed estates in early eighteenth-century Virginia.[50] Heir William Beverley served on the important Governor's Council and speculated successfully on land in the Valley of Virginia. He married Elizabeth Bland of Prince George County and around 1730 moved to the 4,000-acre tract in Essex County on which he built the first house that took her family name.[51] This earlier Blandfield, which was long thought to be the present house, must have resembled the large and informal predecessor of Mount Airy. It must also have been well furnished, for William Beverley possessed sufficient belongings to list in his will of 1756 household goods he owned in England.

In 1750 William Beverley took his son Robert (the builder of the present Blandfield) to England to begin a formal education. The father kept a diary of their travels, which must have left an impression on the ten-year-old son. On August 30, they journeyed to Chatsworth and lodged nearby. The next day, as the father recorded, they "Viewed Chatsworth, a noble house."[52]

Young Robert Beverley remained in England through the decade of the 1750s, maturing at Wakefield Grammar School, Trinity College, Cambridge, and the Middle Temple, from which he was called to the bar in 1761.[53] After his return to Virginia and marriage in 1763 to Maria Carter, herself accustomed to the splendid setting of her father's Sabine Hall, Robert Beverley eventually replaced his late father's Blandfield with a house closer in grandeur to those he had known during the previous decade.

After such a long and apparently enjoyable sojourn in England, Robert Beverley's sentiments were directly opposed to those of such patriots as his father-in-law, who referred to him on at least one occasion in 1775 as "the conceited wise one."[54] When in 1787 this former Tory (of course not the only one in Virginia) was reappointed magistrate for Essex County, one disgruntled neighbor complained to the governor that Beverley was unfit because "he associated during the war *only* with men of sentiments notoriously inimical to the cause of America," he was "averse to our present Government," and he had even "refused toasting Gen'l Washington and the American army during the war."[55] The family recovered from this temporary crisis and continued to prosper, so that by the time of the Civil War owner William Beverley's estate was spoken of by a cousin as worth "about a million & a half of dollars."[56]

By late 1769 Robert Beverley had become sufficiently displeased with his father's fifty-year-old house to complain in a letter to father-in-law Landon Carter, one of the great hosts of the Northern Neck, that "I am perfectly disengaged from all Balls & Entertainments, & am retired into a snug Corner of my old Cabin."[57] Apparently Beverley's discontent had driven him to action earlier than this letter, for in February of the next year, Carter viewed the foundations of a new Blandfield and made note in his diary of what he saw. He betrays the interest and involvement of them both in architecture:

[I] rode to see Beverley's building just raised to the surface of the earth. I believe it will be [torn]omly but in order to make it so I wish both the fronts are not spoiled for I find there is to be a building at each end which goes no higher than the first story. As I understood Mr. Beverley, in order to get some range in the funnels of his Chimneys, he intends to carry one of them over the ceiling of the roof to the main funnel. I endeavoured to advise him against this, being satisfied of the danger if not of the certainly of its smoking but I suppose a plan laid will lead him at all hazzards into the resolution of doing it.[58]

If unexecuted and unclear, Beverley's proposed scheme for chimneys was certainly innovative. About 1770 he must have written to a Liverpool merchant seeking an indentured craftsman to realize his interiors: "As I propose building an House and doing it in the most easy Terms, I have taken this opportunity to [ask you] to procure me an House joiner [of] a good character both as a workman and a well behaved man."[59] We have no information as to whether the carpenter he did secure was found in England, or was well behaved.

The floor plan (fig. 9) that Landon Carter viewed in its foundations in 1770 is of the type seen in plates 42 and 65 (fig. 10) in James Gibbs's *Book of Architecture*, the source used at Mount Airy.[60] In plates 42, 39, and 64 a central "Hall" leads to a central "Dining-Room" of similar dimensions. The stairs are to the sides, entered from the hall, and beyond the dining room is the garden. The corner apartments in these several plates are vaguely described as "Withdrawing-Room," "Bed-chamber," "large Room," "Dressing Room," "Waiting Room," and "Library"—perhaps a clue to the use of so many rooms by Robert Beverley, who must have been equally indecisive regarding function. In elevation, plates 63 and 64 by Gibbs may be very close to the original disposition of Blandfield without its large nineteenth-century porches.

In Robert Beverley's 1795 insurance policy with the Mutual Assurance Society, no porches are noted. But in the 1805 and 1816 policies taken by his son, a "Wooden Portico 2 stories high, 25 × 10 ft." is recorded on both fronts of the house, which were possibly transformed to vaguely resemble a classical temple, following the precedent of Thomas Jefferson's Capitol in Richmond. This occurred at Sabine Hall in the 1820s. A third and final arrangement at Blandfield was decided upon in 1847, when the Van Ness joiners from Washington, used by cousin William H. Tayloe at Mount Airy the previous year, were engaged by William B. Beverley of Blandfield. The $2,600 fee was the same as at Mount Airy, with the addition of another $500 "to build two porticos ten by twenty four feet with fluted columns, . . . a Deck roof & a railing of Iron or wood or both thereon."[61]

On the exterior the Van Nesses also installed new window frames, sash, and glass. They "repaired" the double-hip roof, apparently replacing cypress or cedar shingles with slate. Only paint was needed for the modillioned cornice, which, as Palladio would say, encompasses the whole house like a crown. Of course, the major feature of the Blandfield facades is the mid-Georgian central pavilion and pediment, which gives Palladian emphasis at the center. In two respects the brickwork at Blandfield is unusual: as at neighboring Elmwood flat arches over the windows splay at an extreme angle of forty-five degrees, and the brick used for the connecting passages and dependencies is slightly smaller than the brick that composes the center house.[62] As Robert Beverley tried to explain to Landon Carter, the wings are low, tending even to sink partially below view from the river ascent (fig. 11). Robert Beverley's grandfather had written that such dependencies on the Virginia plantation separated the "Drudgeries" of work "from the Dwelling Houses, which by this means were kept more cool and Sweet."[63] The northwest wing housed the kitchen, while opposite was what the turn-of-the-century insurance appraisers labeled the "Office." It probably always provided living space as well.

The 1847 interior renovation of Blandfield called for "all the wainskirting" of "the principal story" to be "take[n] out," an action that caused considerable confusion after the Van Nesses were forgotten. In 1909 Edith Sale attributed the loss to Federal troops: "the hand-made panelling and wainscoting," she wrote, "were wickedly torn from the walls to be burnt or carried away."[64] With the rediscovery of the Van Ness document, blame shifted to owner William B. Beverley, who had increased the family fortune through cotton operations in Alabama. His pursuit of fashion led to the modern quip that the fire at Mount Airy caused the destruction of two interiors. But the original "wainskirting," though it must have added a sense of scale noticeably absent today, would not have been full-length (floor-to-ceiling) in a mid-Georgian house. And two surviving letters from builder Robert Beverley indicate that initially the interior design of Blandfield featured fashionable wallpapers which after seventy years of use may have sufficiently deteriorated to have warranted replacement.

In April 1771 Beverley wrote to London agent Samuel Athaws regarding wallpaper: "I have been some Time employed in building an House, & as I am desirous of fitting it up in a plain neat Manner, I w[oul]d willingly consult the present Fashion, for you know that foolish Passion had made its way, Even into this remote Region. I observed that L[or]d B. had hung a room with plain blue Paper & border'd it with a narrow stripe of gilt Leather, w[hi]ch I thought had a pretty effect."[65] "Lord B." was Governor Botetourt, who had just papered the ballroom of the Palace in Williamsburg in a manner that may have been copied to some extent by Beverley in his spacious double hall, his new area for entertaining. Three months later he ordered from Athaws wallpapers that were "Pea green flowered, . . . yellow, . . . Stucco color large patterns of pillars and galleries, . . . The borders to be of paper." Also listed on the invoice are "white lead, light stone color, bright olive, dark chocolate, and whiting."[66] Whether all of this was paper or some of it paint, clearly Beverley's house never had full paneling as was popular in the colonies early in the century.

The great halls at Blandfield, with stairs removed to each side, are large in every dimension for the purposes of grandeur and coolness in the summer, the latter an old Virginia tradition noted by the builder's grandfather.[67] Here was space for entertaining that Beverley's neighbors must have envied. As Virginians used their halls for summer dining, and as Gibbs's *Book of Architecture* offered plates of a similar English plan where one hall leads to a dining room, it seems likely that Robert Beverley would have put his second hall to that use. Because each of these important center rooms is heated by a fireplace, both could function throughout the year.[68] Thus the four corner chambers could be equal in size, free for almost any function that interested Beverley.

The present dining room to the southwest may initially have served for less formal meals, for it and that to the northwest are closest to the kitchen. In these rooms remain "Chimney Pieces of the Grey Marble, perfectly plain" that Beverley requested from London in April 1771 after seeing "several . . . imported here [in] late Years." "I am told they are cheap," he wrote, "& I think them infinitely neater than the Embossed or figured ones." The enormous quantity of silver inventoried in the house when Robert Beverley died in 1800 probably was kept in one or both of these western rooms.[69] Silver tea "urns," large and small silver "waiters," silver coffee, tea, and cream "potts," silver "Castors," candlesticks, and dozens of silver spoons of various types comprised only a part of a large service that must have occupied considerable space, perhaps even an entire room as in Williamsburg at the Governor's Palace. The library of "538 Volumes of Books" valued at $270 and said to have been begun by the builder's grandfather must have occupied one of these lower rooms.

As the 1800 inventory lists "85 chairs through the house," in addition to "2 large chairs," "4 armed chairs," and "6 mahogany Chairs"—a total of ninety-seven chairs—a lower southeast room may have always functioned as a secondary parlor, or "withdrawing room" as James Gibbs might have labeled it. In the same way, as there were "13 beds complete" at Blandfield appraised for the huge amount of $910, the northeast lower bedchamber, the only ground-floor room with a closet, was perhaps the builder's private chamber. If so, he would have enjoyed its coolness in the summer, as did his father-in-law on the ground floor at Sabine Hall. In 1800 expensive carpets covered the floors of important rooms. The "2 Turkey Carpets" ($140) and "1 Floor cloth complete with side pieces" ($50) were likely among many found on the first floor.

The closet in the northeast lower chamber seems to be atypical of the initial interior of Blandfield, which was furnished in 1800 with a number of case pieces, including six "presses" and "24 Trunks." One was the clothespress still at Blandfield, which is seemingly unique in its combination of a double-door bookcase section above a clothespress. This and a pair of bookcases, unusual in their arrangement of interior shelves, were products of Williamsburg craftsman Peter Scott.[70]

Much of the storage furniture, of course, would have been used upstairs in six ample bedchambers placed alongside a cross hall stretching from staircase to staircase. The "Dressing tables," "looking glasses," bedsteads, and chairs of the

1800 inventory furnished these rooms, which were simply trimmed and plastered by the Van Nesses.

In the cellars below Robert Beverley stored his wine, including five pipes of Madeira, an enormous quantity totaling perhaps as much as 750 gallons or several thousand bottles. In the tradition of his grandfather who yearly produced this much of his own wine, this Robert Beverley also lived well. At his death the Blandfield estate was appraised at $60,641, while his plantations in neighboring counties earlier were valued at over an additional £15,000.

The earliest writers on Blandfield noted a small garden nestled between the dependencies on the land front of the house extending to the south (the orientation recommended in garden books) to a slight drop in the level of the lawn. To the northeast, where we might expect a larger formal garden as part of a main approach from the river, there remains only a large and rough lawn probably cleared during the Georgian period to resemble a "bowling green." If Gibbs can be valued as a source for Blandfield, the main entrance may always have been from the land side, leading to a hall and dining room, with a garden beyond or, just as likely, on level land to the sides.

Blandfield and Mount Airy stand as very tangible evidence of a colonial heritage perhaps best marked by man's confidence in finding solutions. Not the least of colonial Virginian's achievements involved the art of building, where he often met with success by means of Palladian schemes developed earlier in England. He had only to turn to an established vocabulary and rules of procedure to create designs that are distinguished from the earliest colonial attempts by their harmony and beauty. Palladio would explain that the forms and plans he defined and the colonists modified were ideal, which is probably true since they still seem that way today. Blandfield and Mount Airy testify to the validity of his architectural theory.

NOTES

The late James Fleming (1924–1982), Librarian at the Virginia Historical Society, assisted me with this research and on numerous other occasions with the same kindness and enthusiasm he offered everyone.

[1] Isaac Weld, *Travels through the States of North America and the Provinces of Upper and Lower Canada during the Years 1795, 1796, and 1797* (London, 1800), 1:146.

[2] Robert Beverley the historian, grandfather of the Robert Beverley who built Blandfield, was visited by John Fontaine in 1715 who was impressed by Beverley's vineyard (which that year reportedly produced four hundred gallons of wine) and recorded, "This man lives well" (Alice B. Lockwood, ed., *Gardens of Colony and State* [New York, 1934], 1:46). In the words written by a contemporary, John Tayloe I, father of the builder of Mount Airy, "live[d] in a very genteel manner and both Tayloe & his lady are as agreeable people as I know" (Catesby Cocke to his sister Mrs. Pratt, Feb. 17, 1724, "Jones Papers," *Virginia Magazine of History and Biography* 26 [1918]:71).

[3] *The Four Books of Andrea Palladio's Architecture*, ed. Isaac Ware (London, 1736), p. 46.

[4] Hugh Morrison, *Early American Architecture* (New York, 1952), p. 355. Mount Airy was indeed a country villa, for its builder owned also a house in Williamsburg and his son a house in Washington.

[5] For genealogy, see "Resignation of John Tayloe from the Council," *Virginia Magazine of History and Biography* 17 (1909): 369–73; William Meade, *Old Churches, Ministers, and Families of Virginia*, 2 vols. (Philadelphia, 1857), 2:181–82); Elizabeth Lowell Ryland, ed., *Richmond County, Virginia: A Review Commemorating the Bicentennial* (Warsaw, Va., 1976), p. 115.

[6] Estate of John Tayloe, Richmond County Will Book 5, 1725–53, Virginia State Library (hereafter VSL). The contents of the house were valued at the large sum of £628.

[7] Winslow M. Watson, comp., *In Memoriam: Benjamin Ogle Tayloe* (Washington, D.C., 1872), p. 347.

[8] Edith Sale, *Interiors of Virginia House of Colonial Times* (Richmond, 1927), p. 149.

[9] Arthur Brooke, "A Colonial Mansion of Virginia," *Architectural Review* 6 (1899): 93.

[10] "Resignation of John Tayloe," *Virginia Magazine of History and Biography* 17 (1909): 371.

[11]June 9, 1754, Letterbook of Edmund Jenings, Virginia Historical Society (hereafter VHS). I am indebted to Gwynne Tayloe for directing me to this letter and to Howson Cole, Curator of Manuscripts at the Virginia Historical Society, for helping me dicipher its barely legible passages.

[12]Oct. 27, 1759, Account Book of John Tayloe, 1749–68, VHS.

[13]James Gibbs, *A Book of Architecture* (London, 1728), Introduction, p. i.

[14]Palladio, *The Four Books*, ed. Ware, p. 47.

[15]Thomas Waterman strongly argued a generation ago in his book *The Mansions of Virginia, 1706–1776* (Chapel Hill, N.C., 1946), p. 256, that the land facade of Mount Airy derives from the earl of Aberdeen's Haddo House in northern Scotland, pictured in an obscure volume by William Adam (father of better-known Robert Adam) entitled *Vitruvius Scoticus*, thought to have been published c. 1750. But the similarities are limited and certainly not unexpected, and recent research on *Vitruvius Scoticus* shows the book may have been conceived and begun around 1726, but was not published until c. 1812, long after it could have influenced the architecture of the English colonies. See the introduction in William Adam, *Vitruvius Scoticus*, ed. James Simpson (New York and Edinburgh, 1980).

[16]One person, presumably a slave, who may have helped construct the stone walls of Mount Airy was referred to as "Colonel Tayloe's stonecutter, Ralph" by Landon Carter, who borrowed him on more than one occasion (*The Diary of Colonel Landon Carter of Sabine Hall, 1752–1778*, ed. Jack P. Greene [Charlottesville, Va., 1965] p. 259).

[17]Many questions on the original exterior have not been yet satisfactorily answered. For more related information, see Feb. 12, 1829, Tayloe Diary, VHS; William Rasmussen, "Sabine Hall, A Classical Villa in Virginia" (Ph.D. diss., University of Delaware, 1979), p. 224; *Journal & Letters of Philip Vickers Fithian, 1773–1774: A Plantation Tutor of the Old Dominion*, ed. H. D. Farish (Williamsburg, Va., 1945), p. 106. The question might be settled more conclusively by paint analysis of a sample from either of these dependencies.

[18]Fithian, *Journal*, p. 106.

[19]Plate 59. See also plates 37 and 61.

[20]Testimony of John Tayloe Lomax, "a witness called by the plaintiff," 1845, Tayloe Papers, VHS.

[21]Articles of agreement, Aug. 31, 1846, ibid. This document indicates that some of the ground-floor doors (made of walnut) possibly were saved from the fire.

[22] For contrast, according to the lithograph, the basement level and the connecting units at Mount Airy seem to have been painted white but were not stuccoed first like the more important facades.

[23] Fithian, *Journal*, p. 126.

[24] Mutual Assurance Society Policy, Sept. 18, 1805, VSL. In 1899 Arthur Brooke labeled this building the little school and nursery for the children ("A Colonial Mansion of Virginia," p. 93).

[25] Possibly William Salmon, *Palladio Londinensis* (London, 1734), pl. 27.

[26] In the *American Turf Register and Sporting Magazine* 6, no. 5 (1835): 209, is a portrait of Tychicus, a thoroughbred horse of William H. Tayloe, "loose in the Park, at Mount Airy, his owner's residence." Off in the background, through his legs, can barely be seen the house: it seems to be white in color, a crest is in the pediment, the recessed entrance way is arched, and the fence with urns to the left is balanced by an identical arrangement to the right that terminates in a building one-half the size of the kitchen. This was either an office (according to Brooke) or the dairy (according to the insurance policies). This depiction of the house could have been merely borrowed from the Pendleton lithograph. Fithian tells us of the Mount Airy exterior only that it was "finished curiously," a reference probably to the quoins, belt courses, and pavilions. (*Journal*, p. 126).

[27] Weld, *Travels*, 1:56.

[28] Carter, *Diary*, pp. 533, 907.

[29] Brooke, "A Colonial Mansion," p. 94.

[30] Elizabeth Brand Monroe, "William Buckland in the Northern Neck" (M.A. thesis, University of Virginia, 1969), pp. 19, 39–41; Tayloe's Account Book, VHS. On Oct. 16, 1762, Tayloe in Williamsburg asked neighbor Landon Carter to "ride to Mt. Airy sometimes & give . . . friendly hints of admonition to Mr. Buckland" (Sabine Hall Collection, University of Virginia Library).

[31] John Tayloe II mentioned in a letter of Jan. 3, 1768, "8 chairs and 2 elbow ones that are in Buckland's hands to sell" (Monroe, "Buckland," p. 23).

[32] False windows abound on the second level on the two side elevations: the four corner windows and the eastern section of the southeastern Palladian window do not penetrate to the interior.

[33] Brooke, "A Colonial Mansion," p. 94; F. C. Baldwin, "Early Architecture of the Valley of the Rappahannock: Mount Airy," *Journal of the American Institute of Architects* 4 (1916): 453.

[34] Fithian, *Journal*, p. 126.

[35] Richmond County Personal Property Tax Books, VSL.

[36] Wallace Gusler, *Furniture of Williamsburg and Eastern Virginia, 1710–1790* (Richmond, Va., 1979), identifies the chairs and dropleaf tables as Williamsburg products.

[37] The small dining room at Sabine Hall is listed in an inventory while that at Nomini Hall was noted by the tutor Fithian.

[38] "Samuel Ripley and the Tayloes," *Tyler's Quarterly Historical and Genealogical Magazine* 6 (1924): 4.

[39] Watson, *Tayloe*, p. 246.

[40] Brooke, "A Colonial Mansion," p. 94.

[41] Baldwin, "Mount Airy," p. 453.

[42] Richmond County Personal Property Tax Books, VSL.

[43] Watson, *Tayloe*, p. 246.

[44] *The Writings of George Washington*, ed. John C. Fitzpatrick (Washington, D.C., 1931–44), 25:355–356, 34:74.

[45] November 1760, for £2 8s., Tayloe Account Book, 1749–68, VHS. The first two editions of Miller's *Gardener's Dictionary* were published in London in 1731 and 1741.

[46] Brooke, "A Colonial Mansion," p. 93; Fithian, *Journal*, p. 126.

[47] Watson, *Tayloe*, p. 246; Baldwin, "Mount Airy," p. 454.

[48] Brooke, "A Colonial Mansion," p. 93.

[49] Weld, *Travels*, 1:170.

[50] "Robert Beverley, the Historian of Virginia," *Virginia Magazine of History and Biography* 36 (1928): 333–41.

[51] Edith Sale, *Manors of Virginia in Colonial Times* (Philadelphia, 1909), p. 73.

[52] While in England William Beverley purchased a modest "mahog[any] chest of Drawers" (£4) and "12 chairs," perhaps some of the goods referred to in his will ("Diary of William Beverley of Blandfield during a Visit to England, 1750," *Virginia Magazine of History and Biography* 36 [1928]: 161, 165, 167).

[53] Ibid., p. 27.

[54] Carter, *Diary*, p. 934.

[55] *Calendar of Virginia State Papers*, 4:338–40.

[56] "The Beverley Family," *Virginia Magazine of History and Biography* 21 (1913): 214.

[57] Beverley to Carter, Dec. 30, 1769, VHS.

[58] Carter, *Diary*, p. 362.

[59] C. B. Coulter, Jr., "The Import Trade of Colonial Virginia," *William and Mary Quarterly*, 3d ser., 2 (1945): 312.

[60] It is found also in Colin Campbell, *Vitruvius Britannicus* (London, c. 1715–25), 1: plate 80, probably a less accessible source. See also Gibbs, plates 37, 63.

[61] Articles of agreement, June 17, 1847, Beverley Papers, VHS. In an earlier letter from his cotton base in Alabama, Beverley complained to them about the previous giant portico on the river front that "exclude[d] all air from the cellar adjoining," and he made reference to "the fancy railing as you propose of iron," which we see today (William B. Beverley to G. H. and W. P. Van Ness, April 25, 1847, VHS).

[62] Thomas T. Waterman and James A. Barrows, *Domestic Colonial Architecture of Tidewater Virginia* (New York, 1932), p. 142.

[63] Robert Beverley, *The History and Present State of Virginia,* (1705) ed. L. B. Wright (Chapel Hill, N.C., 1947), p. 290.

[64] Sale, *Manors of Virginia*, p. 74.

[65] Waterman and Barrows, *Domestic Colonial Architecture*, p. 143.

[66] Waterman, *Mansions of Virginia*, p. 265.

[67] Beverley, *History of Virginia*, p. 289.

[68] In the articles of agreement of 1847, the Van Nesses agreed to "cut four doors through the walls [of the halls] so as to open communication from the rooms to the saloons if the said Beverley requires it, & if he does not require it, the said G. H. & W. P. Van Ness is to repair all the fire places & hearths." Apparently the latter was performed.

[69] Appraisal of the estate of Robert Beverley, June 19, 1800, Essex County Will Book no. 16, pp. 15–36, VSL.

[70] Gusler, *Furniture*, pp. 27, 50–51.

1. *Mount Airy, Richmond County, c. 1754–64, land (northeast) facade. (Photograph by Ronald Jennings)*

2. *Mount Airy, river (southwest) façade. (Photograph by Dennis McWaters)*

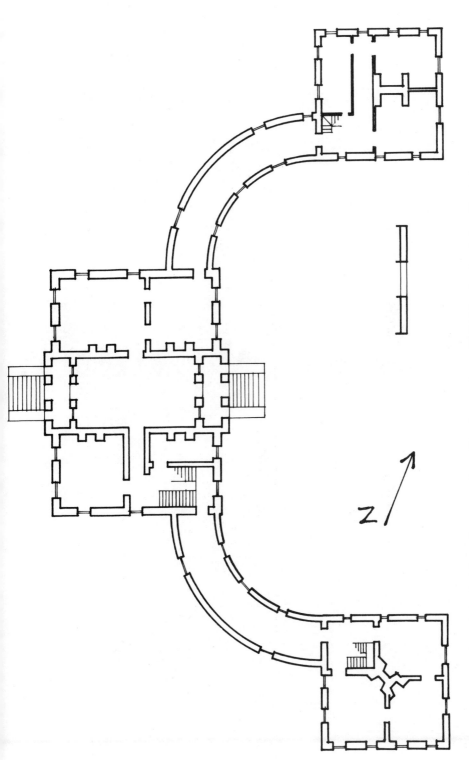

3. *Mount Airy, hypothetical eighteenth-century plan. (Based on a plan published in 1899 by Arthur Brooke)*

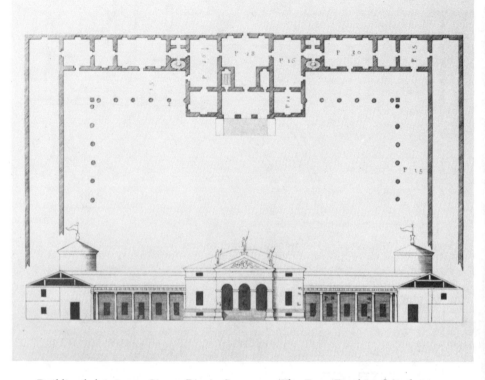

4. *Building belonging to Signor Biagio Sarraceno,* The Four Books of Andrea Palladio's Architecture, *ed. Isaac Ware (London, 1736), bk. 2, pl. 39.*

5. *Mount Airy, river (southwest) facade. (Photograph by Ronald Jennings)*

6. *Mount Airy, lithograph published by Pendleton & Company, Boston, before 1844. (Virginia Historical Society)*

7.   Blandfield, Essex County, c. 1769–70, land (southwest) facade. (Photograph by author)

8. *Blandfield, land (southwest) facade. (Photograph by Dennis McWaters)*

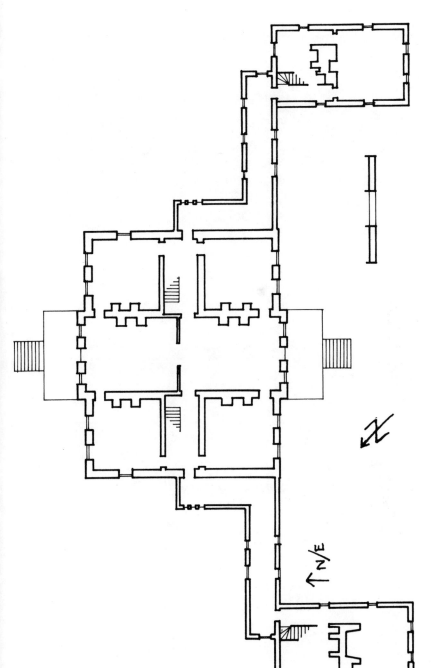

9.   *Blandfield, present plan.*

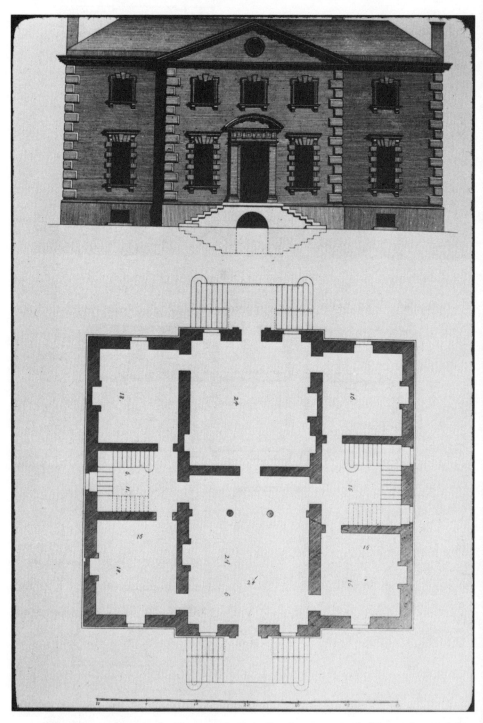

10. *"The Plan and Front of a House made for a Gentleman in Oxfordshire," James Gibbs,*
A Book of Architecture, *pl. 65.*

11. *Blandfield, northeast (river) facade. (Photograph by author)*

CONTRIBUTORS

LIONELLO PUPPI teaches at the University of Padova where he is Director of the Center for Studies in History and Art. He has been a reader at the School of Architecture in Venice, a Fellow of Harvard University, a lecturer at the University of Virginia, and a Fellow of Dumbarton Oaks Center in Washington. Professor Puppi has written extensively on architecture and has published biographies of Palladio, Sanmicheli, Coducci, El Greco, Rembrandt, and Canaletto. Copies of his chapter in Italian are available upon request from the Center for Palladian Studies in America.

CALDER LOTH received his undergraduate and graduate degrees in architectural history from the University of Virginia. He has been employed by the Virginia Historic Landmarks Commission since 1968 and now holds the position of Senior Architectural Historian. He is coauthor of *The Only Proper Style: Gothic Architecture in America*. His most recent project was the preparation of an architectural inventory for the Crown Colony of Gibraltar.

WILLIAM B. O'NEAL, FAIA, professor emeritus at the University of Virginia, was the founder and the first chairman of the Division of Architectural History at the university's School of Architecture. Since his retirement he has devoted himself to writing and lecturing. Recently he returned from England where he collected information on buildings that had influenced the design of the subject of his current work-in-progress, William Lawrence Bottomley.

WILLIAM RASMUSSEN is Coordinator of Education Services at the Virginia Museum of Fine Arts. For the 1981–82 academic year he was on leave from the museum to teach art history at Washington and Lee University. There he conducted research on Mount Airy and Blandfield. His training is in American art, but it is a special interest in the arts in Virginia that underlies his projects in architectural research and educational programming.

CENTER FOR
PALLADIAN STUDIES
IN AMERICA

Andrea Palladio, one of the great masters of the Renaissance culture, has had enduring influence on American architecture, mainly through the *Four Books of Architecture*, which he published in 1570. The Center for Palladian Studies in America has been established in Charlottesville, Virginia, for the purpose of understanding and recognizing Andrea Palladio and "Palladianism" in this country.

The Center for Palladian Studies in America is a nonprofit organization, incorporated under the laws of the Commonwealth of Virginia. It seeks to define "Palladianism," particularly in an American context, to discover American "Palladian" buildings, and to present these findings to the public. These goals are to be achieved through seminars, lectures, scholarly studies, and publications.

Membership to the Center is open to all; fees are $30.00 for regular members, $5.00 for students, and $50.00 for family membership. Contributions, beyond membership fees, are accepted. Both membership fees and contributions are tax deductible.

The Directors and Editor of the Center for Palladian Studies in America invite critical articles, prepared in conformity with *The Chicago Manual of Style*, for presentation at the annual meeting or for publication by the Center. Manuscripts should be submitted in two legible copies. They should be addressed to the Editor, Center for Palladian Studies in America, and be accompanied by a large self-addressed, stamped envelope.

All correspondence concerning manuscripts, membership, and other business should be sent to the Center for Palladian Studies in America, P.O. Box 5643, Charlottesville, Virginia 22905.

CENTER FOR PALLADIAN STUDIES IN AMERICA

DIRECTORS

Howard Burns
Warren J. Cox
Joseph F. Johnston
Mrs. William McC. Blair, Jr.
Ellen V. Nash
Frederick D. Nichols
Lionello Puppi
Edmund A. Rennolds
Dean of the School of Architecture, University of Virginia
Mario di Valmarana
Christopher C. Weeks
Stanley Woodward

OFFICERS

Stanley Woodward, President
Mario di Valmarana, Vice President
Edmund A. Rennolds, Treasurer
Ellen V. Nash, Secretary